# LABORATORY MANUAL FOR

# BIOLOGY OF HUMANS

CONCEPTS, APPLICATIONS, AND ISSUES

GOODENOUGH          MCGUIRE          WALLACE

## MIMI BRES • ARNOLD WEISSHAAR
PRINCE GEORGE'S COMMUNITY COLLEGE

Upper Saddle River, New Jersey 07458

Editor-in-Chief, Science: John Challice
Executive Editor: Gary Carlson
Project Manager: Crissy Dudonis
Executive Managing Editor: Kathleen Schiaparelli
Assistant Managing Editor: Becca Richter
Production Editor: Rhonda Aversa
Supplement Cover Manager: TJS Concepts Inc.
Supplement Cover Designer: James Enering
Manufacturing Buyer: Alan Fischer
Cover Image Credits: Father Teaching Son Basketball – Jim Erickson/Corbis; Spiraling Roller Coaster – Robert Landau/Corbis; Beach Volleyball – Michael Kevin Daly/Corbis; American Gymnast – Wilson Competes Corbis/Bettman

© 2005 Pearson Education, Inc.
Pearson Prentice Hall
Pearson Education, Inc.
Upper Saddle River, NJ 07458

All rights reserved. No part of this book may be reproduced in any form or by any means, without permission in writing from the publisher.

Pearson Prentice Hall™ is a trademark of Pearson Education, Inc.

The author and publisher of this book have used their best efforts in preparing this book. These efforts include the development, research, and testing of the theories and programs to determine their effectiveness. The author and publisher make no warranty of any kind, expressed or implied, with regard to these programs or the documentation contained in this book. The author and publisher shall not be liable in any event for incidental or consequential damages in connection with, or arising out of, the furnishing, performance, or use of these programs.

> **This work is protected by United States copyright laws and is provided solely for teaching courses and assessing student learning. Dissemination or sale of any part of this work (including on the World Wide Web) will destroy the integrity of the work and is not permitted. The work and materials from it should never be made available except by instructors using the accompanying text in their classes. All recipients of this work are expected to abide by these restrictions and to honor the intended pedagogical purposes and the needs of other instructors who rely on these materials.**

Printed in the United States of America

10 9 8 7 6 5 4 3 2 1

ISBN 0-13-144960-5

Pearson Education Ltd., *London*
Pearson Education Australia Pty. Ltd., *Sydney*
Pearson Education Singapore, Pte. Ltd.
Pearson Education North Asia Ltd., *Hong Kong*
Pearson Education Canada, Inc., *Toronto*
Pearson Educación de Mexico, S.A. de C.V.
Pearson Education—Japan, *Tokyo*
Pearson Education Malaysia, Pte. Ltd.

# Contents

|  |  | Preface | iv |
|---|---|---|---|
| Exercise | 1 | Introduction to the Scientific Method | 1 |
| Exercise | 2 | Using the Compound and Dissecting Microscopes | 21 |
| Exercise | 3 | Functions and Properties of Cells | 39 |
| Exercise | 4 | Investigating Cellular Respiration | 59 |
| Exercise | 5 | The Skin: An Example of an Organ System | 79 |
| Exercise | 6 | The Musculoskeletal System | 97 |
| Exercise | 7 | Examination of Skeletal Structure | 117 |
| Exercise | 8 | The Nervous System | 137 |
| Exercise | 9 | Introduction to Forensic Biology | 153 |
| Exercise | 10 | The Circulatory System | 171 |
| Exercise | 11 | Introduction to Anatomy: Dissecting the Fetal Pig | 189 |
| Exercise | 12 | Organ Systems in the Abdominal Cavity | 209 |
| Exercise | 13 | Nutrient Analysis of Foods and Choices for Good Health | 233 |
| Exercise | 14 | The Reproductive System | 251 |
| Exercise | 15 | Mitosis and Asexual Reproduction | 273 |
| Exercise | 16 | Connecting Meiosis and Genetics | 289 |
| Exercise | 17 | Useful Applications of Genetics | 317 |
| Exercise | 18 | Introduction to Molecular Genetics | 335 |
| Exercise | 19 | Biotechnology: DNA Analysis | 359 |
| Exercise | 20 | Using Biotechnology to Assess Ecosystem Damage | 385 |
| Appendix | I | Self-Test Answers | 399 |

# Preface

**Understanding Human Biology: Laboratory Exercises** is designed for a one-semester, nonmajors, general biology course with a human focus. The topics and exercises are general enough to be compatible with most introductory level human and general biology texts currently in use. The activities demonstrate that basic biological concepts can be applied to a wide variety of plants, animals, and microorganisms.

This book is unique not because of the specific topics covered, but because of the approach to these topics. These laboratory exercises are planned to help your students

- gain practical experience that will help them understand lecture concepts,
- acquire the basic knowledge needed to make informed decisions about biological questions that arise in everyday life,
- develop the problem-solving skills that will lead to success in school and in a competitive job market,
- learn to work effectively and productively as a member of a team.

The basis of scientific work is asking questions and answering them by observations or experiments. Thus, we feel strongly that the most important goal of an introductory course in human biology should not be to simply outline facts about anatomy, physiology, or new techniques. Rather, students should come away with some understanding of the processes of investigation that are basic to science, and how scientists work to solve problems.

We hope that working through this laboratory manual will be an exciting experience for both students and instructors, and one that will leave students, better prepared to meet the demands of our increasingly scientific society.

## SPECIAL FEATURES FOR YOU, THE INSTRUCTOR

### Three-Pronged Approach to Laboratory Learning

- **Elicit Interest.** Students perform better when they have a clear understanding of the reason and objectives behind each exercise. Students are often more willing to proceed if they know why they are asked to do each activity and what objectives they are expected to accomplish during the laboratory period.
- **Clear Directions.** Students need clearly and concisely written directions of how to proceed. This is the first and last science class for many students, and we can make very few assumptions about lab skills they might bring into the class.
- **Establish Relevance.** A well-designed lab needs to help students understand the meaning and importance of the findings they just gathered, and to establish relevance between laboratory and lecture topics and everyday life.

# Preface

*Clarity*

- All content and problem-solving goals are accomplished by using a **simple, non-threatening approach** to the lessons. Even students with little science background are able to **get involved** and master the material.

*Instructional Objectives*

- A clear set of **Instructional Objectives** begins each exercise. This helps students identify the most important concepts that will be emphasized in each activity.

*Focus on Thinking and Problem Solving*

- One of the most important features of this laboratory manual is the **infusion of thinking and problem-solving skills into the content areas**. Research has shown that it is easiest to master thinking and problem-solving skills when thinking and content are learned together. **Laboratory Manual for Biology of Humans** is based on this approach. Activities and questions within an exercise build on previously learned information and encourage students to transfer information from one section of the course to another.

*Team Building Opportunities*

- Most laboratory activities emphasize a **team approach**. Group work is encouraged and often required. In the real world, we are expected to interact with others to solve problems and complete projects. This approach provides opportunities for students to work together, share ideas, and function effectively in groups to accomplish tasks.

*Real-life Connections*

- Questions throughout the activities and those incorporated into the **Self Tests** are designed to connect the laboratory activities to timely subjects and real-life experiences. The exercises are intended to stimulate interest in topics that can help students make decisions regarding their own health and nutrition, understand current topics in the news, and their role in the environment.

*Many Opportunities to Practice Skills and Reinforce Concepts*

- Activities are **reinforced with relevant questions**, many more than can be found in any competing editions. It is not only the number of questions, but also the **type** of questions that are important. Questions are designed to provide **continuous feedback** and **elicit targeted thinking**. This keeps the learner "on track" and fosters development of problem-solving skills.
- Through **questions interwoven** into each activity, **Comprehension Checks**, and **Self Tests**, recall of important concepts is reinforced. Students also have the opportunity to transfer content knowledge gained to new situations.

*These Lessons Work!*

- These laboratory activities have been **fully tested** with thousands of students on this and other campuses around the United States.
- Experience has shown that the simplified structure and easy-to-understand language of this manual works well with a diverse student population.

*Requires Minimal Supplies and Equipment*

- These exercises are cost-effective and affordable even for schools with budget constraints. **Most schools will already have most of the equipment needed to perform the lab activities.**

*Designed for Simplicity*

- All activities are **easy to set up and maintain**. Our program is large and multi-sectional. Despite large numbers of instructors and students, this laboratory curriculum is incorporated smoothly in all sections. The simplicity of the laboratory approach has allowed new personnel to be **"up and running" with minimal preparation time.**
- Since the exercises require minimal equipment, many activities throughout the manual are **suitable for use in online sections**.

*Adaptable to Any Curriculum*

- Each exercise is designed for use in a three-hour laboratory period, but can easily be adapted to accommodate two-hour or 90-minute sessions. To provide **maximum flexibility** for instructors, each exercise is **divided into several activities**. Activities can be deleted or presented as demonstrations without diminishing the value of the remaining components.
- Laboratory exercises can be presented **in sequence or rearranged** to suit your needs. The manual includes 20 exercises, allowing you the latitude to tailor the laboratory curriculum to your course and the equipment available at your institution.

*Support Materials—Instructor's Guide.*

The Instructor's Guide provides you with

- complete preparation and setup instructions for each activity,
- a complete Answer Key for all activity questions,
- helpful hints to facilitate smoothly running laboratories,
- master charts for recording results of classroom experiments.

## SPECIAL FEATURES FOR YOU, THE STUDENT

*Informal Style*

- The text is written in an active voice with easy-to-understand language. Key terms and important definitions are highlighted with bold print for easy recognition. Large spaces are provided throughout the manual for you to record and fully explain your answers.

# Preface

*Active Learning Experience*

- Every exercise gives you an opportunity to be an active participant in the scientific method. You will form hypotheses, set up experiments, collect data, record your data in graphs and charts, and draw conclusions from your experimental results.

*Self-guided Approach*

- The laboratory procedures are clearly outlined to allow you to progress through each activity independently.

*Tools for Success*

The following components of each exercise will help you to succeed:

- **Instructional Objectives.** The objectives are listed first in each exercise so that you will be able to focus your attention on the main concepts of each activity.
- **Content Focus.** Each exercise includes a brief discussion of the background information that you will need to understand the subject of the exercise and prepare you to complete the activities that follow.
- **Notes and Cautions.** Note boxes provide helpful hints for solving problems and accomplishing laboratory tasks. Pay special attention to caution boxes that provide important safety information.
- **Comprehension Checks.** Stop and complete the Comprehension Check questions to get immediate feedback on your understanding of the basic principles covered in each activity. These questions also provide a chance for you to apply what you have learned to situations outside the classroom.
- **Check-off Boxes.** These boxes allow your instructor to check your progress before you move on to a new activity.
- **Self Tests.** Answer these questions after completing the laboratory exercise. Self-Test questions allow you to assess your comprehension and apply your knowledge. Answers to the Self Test questions can be found in Appendix I at the back of the book.

# ACKNOWLEDGMENTS

Any project of this size and scope cannot be completed without support and cooperation from many people. We would like to express our deep appreciation for their contributions.

Thanks to

Louis Renaud, Chairman of the Department of Biological Sciences at Prince George's Community College, for providing the opportunity to do this project and supporting professional development.

Michael Garvey, Federal Bureau of Investigation, Washington, D.C., for providing his time and technical expertise in the area of forensic molecular genetics.

Janet McMillen, Professor of Biology, Prince George's Community College, for reviewing newly developed laboratory activities. Her helpful suggestions have contributed substantially to the effectiveness of these laboratory exercises.

Kenneth Thomulka, Professor of Biology, University of the Sciences in Philadelphia, for allowing us to adapt an imaginative and unique application of biotechnology to the education arena.

Charlotte Weisshaar, for proofreading and editing our many manuscripts, thus ensuring continuity within and between exercises.

Robert Ewing and Margaret Ryan, for the excellent artwork and photographs they contributed to this manual.

the Biology Students of Prince George's Community College, for classroom testing our laboratory exercises and offering suggestions for improvements.

Crissy Dudonis, Project Manager for Biology, Prentice Hall, for lending her skill and support to this project. Her efficient and effective management played a significant role in making this project possible.

## REVIEWERS

John S. Campbell, *Northwest College*
Renee E. Carleton DVM, *Berry College*
Dwight O. Kamback, *Lake City Community College*
Will Kleinelp, *Middlesex County College*
H. Roberta Koepfer, *Queens College, CUNY*
Monica McGee, *University of North Carolina–Wilmington*
William Simcik, *Tomball College*
Jason Yoder, *Itasca Community College*

# EXERCISE 1

# Introduction to the Scientific Method

## Objectives
After completing this exercise, you should be able to:
- use the scientific method to solve problems
- organize information to facilitate analysis of your data
- draw graphs that present data clearly and accurately
- interpret data in tables, charts, and graphs
- draw conclusions that are supported by experimental data
- analyze data using common statistical measures
- apply your knowledge of the scientific method to real-life situations.

## CONTENT FOCUS

What is science? What do scientists *do* all day? These are not easy questions for most of you to answer. The widespread idea of a nerd in a white lab coat does not apply to most scientists. So, what are scientists really like? They all have the "**three Cs**" in common.

Just as you are, scientists are **curious** about the world around them. They ask questions about everything. Can my diet cause heart disease? Why does the river look brown instead of blue? How can squirrels remember where they bury their nuts? Why do some cars get better mileage than others? Science is a method for answering these and many other questions.

Scientists don't accept things without **collecting information**. All the facts relating to a problem or question have to be carefully explored and checked for accuracy.

Scientists are **comfortable with new concepts**. If a better explanation can be found, scientists are not afraid to give up old ideas for new ones.

To make the three Cs happen, scientists have developed a series of steps in investigation called **the scientific method**. Through trial and error, the scientific method has proven to be an efficient and effective way of attacking a problem. You have probably used some version of the scientific method many times in your life—without being aware of the steps you were following.

## ACTIVITY 1 — FORMING HYPOTHESES: DARING TO BE WRONG

*Preparation*

There are several different ways that a problem can come to your attention. Someone may **assign** you the problem (this happens often in a school or a work situation), the problem may **thrust itself** upon you (your car won't start), or you may discover the problem by simply being **curious** about something you have seen.

Let's begin with a simple situation that you might face any day.

> **The Problem**
>
> You drive to school and park in your usual spot. As you walk across the campus after your morning biology class, you discover that you can't find your car keys. You have a problem!

An easy way to attack the problem is to make an **educated guess** about the possible solution to the problem. It is an "educated" guess because you use all the background information that is available when making your guess.

In scientific terms, an educated guess is called a **hypothesis**.

1. On the next page, in Table 1-1, you will find a hypothesis that might shed some light on this problem.

2. **Complete the table** by adding some hypotheses of your own.

**Exercise 1** ■ *Introduction to the Scientific Method* 3

| TABLE 1-1 <br> KEY LOSS: POSSIBLE HYPOTHESES |
|---|
| **A:** The keys are in my book bag. |
| **B:** |
| **C:** |
| **D:** |

Check your hypotheses with your instructor before you continue.

## ACTIVITY 2    FORMING HYPOTHESES TO SOLVE PROBLEMS

### The Problem

**You were absent from chemistry class the day your professor gave out the instructions for making an important solution needed for your laboratory experiment. No problem. Your roommate was in class and had copied down the formula for you. You rush off to chemistry lab and prepare the solution, but when you use it in your experiment, it doesn't perform as expected.**

In **Table 1-2, list three hypotheses** about why the formula didn't work. Don't forget—the hypotheses must be **testable!**

| TABLE 1-2 CHEMISTRY EXPERIMENT: HYPOTHESES |
|---|
| A: |
| B: |
| C: |

 **Check your hypotheses with your instructor before you continue.**

Some hypotheses can be tested by observation only, but more often, you will need a **combination of observation and experimentation** to be sure about the accuracy of your results. To understand how scientists work, you must follow the steps of the scientific method as they are used in actual **experiments. In Activities 3 and 4**, you will **see how scientific method skills** are used to **set up experiments** and analyze the **information (data)** that is collected.

## ACTIVITY 3 — TESTING HYPOTHESES

### The Problem
Investigate the effects of fertilizer on plant growth.

STEP 1:

You form a hypothesis about what you think will happen.

**Hypothesis:** Adding fertilizer will make plants grow taller.

**Exercise 1** ■ *Introduction to the Scientific Method*          **5**

STEP 2:

You design an experiment that compares the growth (in height) of plants that **receive fertilizer** with those grown **without fertilizer. Your design might be similar to the following:**

Begin with **20 plants** (same size, same type), planted in the same-size pots, with the same amount and type of soil, placed on a windowsill with the same exposure to light. All these factors will be held constant.

- An experiment is designed to isolate the factor you are interested in testing. All **other conditions must be held constant**. In this way, you are sure that your **observed results** were caused by the only factor that was varied.

- Since you are investigating the **effect of fertilizer**, you will want to hold all **other factors constant** (plant type, plant size, pot size, amount of water, amount of light, etc.) to avoid confusion. This way you can be **sure** that any differences in height are **due to the presence of fertilizer** and not some other factor.

You decide to measure the growth of your plants (height in centimeters) **once a week for a month**. You will keep detailed **records** of your **observations**.

It is helpful to plan an experiment with a **group** of plants (or animals). There are **two good reasons** to use groups:

- If **unexpected factors** (such as disease) affect one or two experimental subjects, it will not ruin the experiment.

- **Natural genetic variability** will cause some plants to grow taller than others (just as some people grow taller than others). You can separate this effect from that of the fertilizer by measuring the height in a **group** of plants for each treatment (fertilizer and no fertilizer).

You decide that **ten plants will receive identical measured amounts of fertilizer** each week. These are the **experimental** plants. They are receiving the treatment (fertilizer) that will help you test your original hypothesis (does fertilizer make plants grow taller?).

**Ten plants will receive no fertilizer**. These are the **control** plants. They do not receive the experimental treatment. You will use these for **comparison with the experimental group** to help you interpret your results, and to show that any observed differences in height between the two groups are due to the **only difference between them**—application of fertilizer.

In the above example, there are **ten replications of the experimental treatment and 10 replications of the control treatment**.

## Comprehension Check

1. Why is it necessary to divide the plants into two groups (a control group and an experimental group)?

2. Why is it important to keep conditions exactly the same in the control and experimental group, **except** for the application of fertilizer?

3. Why is it better to do the plant/fertilizer experiment with **ten** plants in each group instead of just one or two? Give **two** reasons. **Use your own words**.

   a.

   b.

**Check your answers with your instructor before you continue.**

**Exercise 1** ■ *Introduction to the Scientific Method* 7

# ACTIVITY 4                                    INTERPRETING DATA

Preparation

The month is up. You are ready to draw conclusions from your **data** (the information you have recorded). You will be thinking about what your results mean and whether your hypothesis is **supported**.

The information you collected during your experiment is presented in **Tables 1-3 and 1-4**.

### ✓ Comprehension Check

1. Were there differences in growth between the control and experimental plants?

    If so, which group grew taller? How do you know?

2. Do the results support the original hypothesis? **Explain** your answer.

3. Why is it more accurate to compare the **average** height gain of the control and experimental groups (instead of comparing individual plants)?

**Check your answers with your instructor before you continue.**

### TABLE 1-3
### HEIGHT GAIN (cm) OVER FOUR WEEKS—CONTROL PLANTS

| Plant Number | Initial Height (cm) | Week 1 | Week 2 | Week 3 | Week 4 | Total Height Gain | Avg Height Gain |
|---|---|---|---|---|---|---|---|
| 1 | 10.0 | 1.6 | 2.0 | 3.0 | 2.5 | 9.1 | 2.3 |
| 2 | 11.5 | 2.2 | 1.5 | 1.5 | 2.0 | 7.2 | 1.8 |
| 3 | 9.6 | 1.5 | 2.3 | 2.6 | 2.0 | 8.4 | 2.1 |
| 4 | 9.2 | 2.0 | 3.0 | 2.8 | 1.5 | 9.3 | 2.3 |
| 5 | 10.2 | 2.3 | 1.2 | 1.6 | 2.0 | 7.1 | 1.8 |
| 6 | 11.0 | 3.2 | 1.7 | 2.0 | 3.2 | 10.1 | 2.5 |
| 7 | 10.0 | 2.6 | 3.0 | 3.0 | 1.4 | 10.0 | 2.5 |
| 8 | 9.7 | 4.0 | 2.6 | 4.0 | 2.3 | 12.9 | 3.2 |
| 9 | 10.4 | DIED | – | – | – | – | – |
| 10 | 10.4 | 2.3 | 2.3 | 2.7 | 2.0 | 10.0 | 2.5 |
| TOTAL | | | | | | 84.1 | 21.0 |

### TABLE 1-4
### HEIGHT GAIN (cm) OVER FOUR WEEKS—EXPERIMENTAL PLANTS

| Plant Number | Initial Height (cm) | Week 1 | Week 2 | Week 3 | Week 4 | Total Height Gain | Avg Height Gain |
|---|---|---|---|---|---|---|---|
| 1 | 9.6 | 4.2 | 5.0 | 3.0 | 4.7 | 16.9 | 4.2 |
| 2 | 9.8 | 6.0 | 4.0 | 5.5 | 5.0 | 20.5 | 5.1 |
| 3 | 10.3 | 5.3 | 5.5 | 3.6 | 4.2 | 18.6 | 4.7 |
| 4 | 11.0 | 2.1 | 3.2 | 6.2 | 3.8 | 15.3 | 3.8 |
| 5 | 10.1 | 3.4 | 4.0 | 4.4 | 4.0 | 15.8 | 4.0 |
| 6 | 9.2 | 4.7 | 3.1 | 3.1 | 4.0 | 14.9 | 3.7 |
| 7 | 9.5 | 4.2 | 5.2 | 3.9 | 3.6 | 16.9 | 4.2 |
| 8 | 10.0 | 3.3 | 6.0 | 5.6 | 4.2 | 19.1 | 4.8 |
| 9 | 9.7 | 5.8 | 6.1 | 6.5 | 5.0 | 23.4 | 5.9 |
| 10 | 10.4 | 5.1 | 3.4 | 5.8 | 5.3 | 19.6 | 4.9 |
| TOTAL | | | | | | 181.0 | 45.3 |

**Exercise 1** ■ *Introduction to the Scientific Method*

# ACTIVITY 5    FORMING HYPOTHESES FOR AN EXPERIMENT

*Preparation*

Now that you have had some experience using the steps of the scientific method, you will develop a hypothesis and test it in a simple experiment.

Table 1-6 contains some examples of physical traits and physiological factors that may or may not be related. You can choose to investigate **any two factors from Table 1-6**.

1. **Working alone**, form a **hypothesis** to test in your experiment.

2. **Table 1-5** contains a sample **hypothesis** about the relationship between a **physical trait (height) and a physiological factor** (pulse rate) that are listed in **Table 1-6: Finding Relationships**.

   **Complete the table** by adding your hypothesis.

| TABLE 1-5 HYPOTHESIS ABOUT RELATIONSHIPS ||
|---|---|
| Example: | The taller you are, the faster your pulse rate will be. |
| Your Hypothesis: | |

**Check your hypothesis with your instructor before you continue.**

| TABLE 1-6 FINDING RELATIONSHIPS | |
|---|---|
| CHARACTERISTIC | METHOD OF INVESTIGATION |
| Height | Remove your shoes and stand against the wall. Use a meter stick to measure the distance from the floor to the top of your head (in centimeters). |
| Arm Length | Use a tape measure to measure the length of your arm from your shoulder joint to the tip of your middle finger (in centimeters). |
| Head Circumference | Use a tape measure to measure the distance around your head, just above your ears (in centimeters). |
| Pulse Rate | Place your index and middle fingers on your carotid artery (on either side of the neck). Using a watch that indicates seconds, count the total number of pulse beats for fifteen seconds. Multiply your answers by four to get the pulse rate per minute. Repeat the process twice more and then calculate the **average** number of pulse beats per minute. |
| Shoe Length | Measure the bottom of your shoe from the tip of the toe to the back of the heel (in centimeters). |

## ACTIVITY 6 — DESIGNING THE EXPERIMENT AND COLLECTING DATA

1. **Work in groups of three to four students.**

   Choose **one hypothesis** to investigate from those developed by your group members.

2. **Form a plan** for getting the **data** you will need to test your hypothesis.

   **The following supplies are available** to conduct your experiment: **meter sticks** and **tape measures**.

   Collect your experimental data and record the results in **Table 1-7**.

Exercise 1 ■ *Introduction to the Scientific Method*

| TABLE 1-7 EXPERIMENTAL DATA | | |
|---|---|---|
| SUBJECT | | |
| 1 | | |
| 2 | | |
| 3 | | |
| 4 | | |
| 5 | | |
| 6 | | |
| 7 | | |
| 8 | | |
| 9 | | |
| 10 | | |
| 11 | | |
| 12 | | |

Your fellow students will be happy to volunteer as test subjects in your experiment.

### Hint:

How many pieces of information (data points) should you collect to feel confident about your conclusions? There are probably not enough people in your small laboratory group to make an adequate test sample. How many additional test subjects from the class will you need?

# ACTIVITY 7  ORGANIZING AND SUMMARIZING RESULTS

To determine if there is a relationship between the two factors, **rearrange your data and copy it into the empty table below (Table 1-8)**.

Place the information in order from **smallest to largest** (shortest to tallest, lightest to heaviest, or smallest head to largest head).

### TABLE 1-8
### REARRANGED DATA

| Subject | | |
|---------|---|---|
|         |   |   |
|         |   |   |
|         |   |   |
|         |   |   |
|         |   |   |
|         |   |   |
|         |   |   |
|         |   |   |
|         |   |   |
|         |   |   |
|         |   |   |
|         |   |   |

**Exercise 1** ■ *Introduction to the Scientific Method* 13

# ACTIVITY 8                              GRAPHING YOUR RESULTS

1. Graphs provide a good visual representation of the relationships between the factors investigated in an experiment. **Bar graphs** and **line graphs** are frequently used to present scientific data.

    Figure 1-1 illustrates two different ways to present the same information.

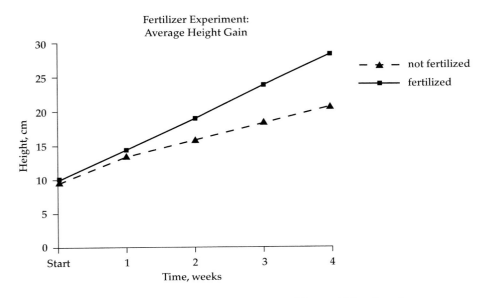

**FIGURE 1-1.** Comparison of Bar and Line Graphs

2. Look at the graphs in **Figure 1-1** and note the following **key points**:

   - The **horizontal** axis is referred to as the **X-axis**. The **vertical** axis is called the **Y-axis**.

     Each axis **must** have a title that clearly explains the numbers listed there.

   - Numbers on the X and Y axes must have an **equal interval** between them (for example, 5, 10, 15 but **not** 5, 10, 20, 50).

   - Lines or bars should be **large and easy to read**.

   - Numbers on the X and Y axes are chosen carefully to make **best use** of the space available.

   - Each graph should have a **title** that describes the subject matter being graphed.

   - It is not permissible to extend lines or bars **outside the margins** of the graph. Adjust the graph scale to make the data fit comfortably.

3. On the graph paper in **Figure 1-2, plot a graph of your experimental results**.

   Plot one characteristic on the **vertical (Y) axis** and the other on the **horizontal (X) axis**.

   Whenever possible, plot the **first characteristic mentioned in your hypothesis** on the **X (horizontal) axis** and the **second** characteristic on the **Y (vertical) axis**.

4. **Discuss the results** with your other group members.

   **Write a conclusion based on your hypothesis and collected data. Support your conclusion** by mentioning facts collected during your experiment.

---

**Did You Know?**

All the information, concepts, and relationships you will read about in your textbook were discovered and verified by the same scientific process you used in this laboratory exercise.

During this course, we will be summarizing hundreds of years of study and experimentation. Information presented on television, in newspapers, and in magazines often has not been confirmed by this same careful process and may not be correct.

# Exercise 1 ■ *Introduction to the Scientific Method*

**FIGURE 1-2.** Experimental Results

# ACTIVITY 9  INTERPRETING DATA USING SIMPLE STATISTICS

Preparation

Often, the data collected in an experiment is in a form that is not easily understandable. Measurements have been established to make it easier to interpret and draw conclusions from large collections of information. We're all familiar with the U.S. Census, which collects huge amounts of information on family size, income, housing conditions, population distributions, and so on.

Simple statistical analysis can reduce the data and convert it to a useable form. A similar approach is used when analyzing the results of large experiments (such as evaluating the effectiveness of new medications or airbags in automobiles). The most commonly used statistical measures are the mean, the mode, the median, the range, and the standard deviation.

>The **mean** is the **average** of a set of numbers. The mean is equal to the **sum of all the numbers in the set divided by the sample size**. For example, to find the average pulse rate of a group of ten students, you would add the pulse values for each of the students together and then divide the answer by the number of students (10).
>
>The mean of a group of numbers often doesn't really give you the information you need to correctly interpret the data. **For example, the following two sets of numbers have exactly the same mean, but the spread (dispersion) of numbers is quite different.**
>
>>**Set 1**: 39, 38, 38, 40, 40  mean = 39
>>
>>**Set 2**: 3, 29, 25, 38, 100  mean = 39
>
>The **mode** is the **most frequently occurring number** in a set. The mode represents the most common response and, therefore, can be used as a prediction to determine market response (for example, which car model will sell the best in a specific area of the United States).
>
>The **median** is the **middle** number of a set when they are arranged in either **ascending or descending order**. If your income level is above the median, for example, your salary is in the upper 50% of salaries being compared. If a set of numbers has no middle value, you can find the median by averaging the middle two numbers in the set.
>
>The **range** is the **difference between the largest and smallest value** in the set. For example, the difference between the number of yards gained by the best and worst running backs in the National Football League.

To demonstrate how applying different statistical measures changes the meaning of results, consider the following set of 20 biology exam scores in **Table 1-9**.

**Exercise 1** ■ *Introduction to the Scientific Method* 17

| TABLE 1-9 BIOLOGY EXAM SCORES |||||
|---|---|---|---|
| Student Number | Score: Exam 1 | Student Number | Score: Exam 1 |
| 1 | 90 | 11 | 88 |
| 2 | 94 | 12 | 54 |
| 3 | 80 | 13 | 32 |
| 4 | 82 | 14 | 47 |
| 5 | 91 | 15 | 25 |
| 6 | 46 | 16 | 56 |
| 7 | 97 | 17 | 59 |
| 8 | 96 | 18 | 60 |
| 9 | 87 | 19 | 87 |
| 10 | 84 | 20 | 86 |

Mean = 72.05

Mode = 87

Median = 83

1. What is the **range** of exam scores in **Table 1-9**?

2. If your **exam score was 80**, was your score in the top 50% of the class? **Explain your answer**.

3. In this situation, is the **mean** a good representation of the class scores? **Why or why not?**

4. While watching television last night, you saw an advertisement for Lose-Fast Weight Control Pills. The 12 women shown lost an **average** of 30 pounds while taking the pills. What additional **statistical** information would you need to make an informed decision whether or not to purchase this product?

**Check your answers with your instructor before you continue.**

## Self Test

For each sentence below, enter the letter of the correct step of the scientific method.

**a. Test hypothesis (by experiment or observation)**

**b. State hypothesis**

**c. State results (facts only)**

1. \_\_\_\_\_ Seeds will grow faster if you fertilize the ground before you plant them.

2. \_\_\_\_\_ In an experiment, 70 of 80 household cockroaches were attracted to peanut butter.

3. \_\_\_\_\_ Tanya grew bacteria from her mouth on special plates in the laboratory. She placed drops of different mouthwashes on each plate.

4. \_\_\_\_\_ Kevin designed a survey to determine how many of his classmates had dimples on their chins and how many did not.

5. \_\_\_\_\_ Plants grown under red light will grow faster than those under white light.

6. \_\_\_\_\_ If acid rain affects plants in a particular lake, it might also affect small animals that live in the same water.

7. \_\_\_\_\_ Maria's experiment showed that chicken eggshells are more resistant to crushing when the hens are fed extra calcium.

Identify the graphing mistakes in **Figures 1-3, 1-4, and 1-5**:

8.

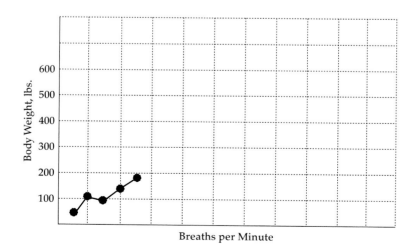

**FIGURE 1-3.** Sample Graph One

**Exercise 1** ■ *Introduction to the Scientific Method*

9.

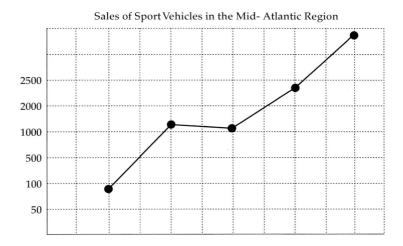

**FIGURE 1-4.** Sample Graph Two

10.

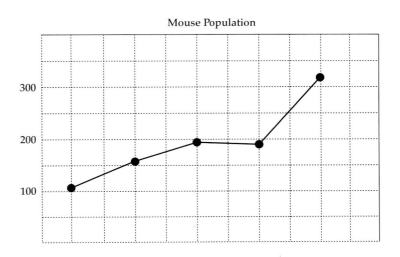

**FIGURE 1-5.** Sample Graph Three

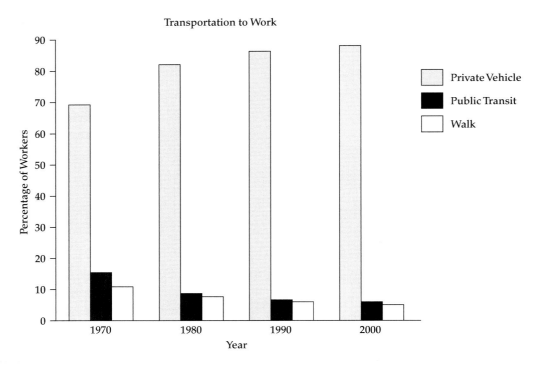

**FIGURE 1-6.** Percentage of Workers Using Various Modes of Transportation

Answer the following questions in reference to the graph in **Figure 1-6**.

11. (**Circle one answer**.) Between 1970 and 2000, the percent of workers who walked to work **increased/decreased/remained the same**.

12. Between 1970 and 1980, the number of people who drove cars to work increased by _____%.

13. In 2000, what percent of workers drove cars to work? _____.

14. The number of workers who used public transportation dropped by 7% between 1970 and _____.

# EXERCISE 2

# Using the Compound and Dissecting Microscopes

## Objectives

After completing this exercise, you should be able to:
- identify the parts of the compound and dissecting microscopes and explain their functions
- choose the correct type of microscope for viewing different specimens
- focus the compound microscope using the scanning, low-, and high-power lenses
- prepare a wet mount slide
- correct viewing problems that commonly occur when using the compound microscope
- accurately describe specimens viewed through the dissecting and compound microscopes
- use the microscope to test a hypothesis.

## CONTENT FOCUS

All organisms, large and small, make valuable contributions to ecosystem function. Many plants and animals are large and easy to see, but many important organisms are too small to be seen without the assistance of magnifying lenses.

You will have an opportunity to get a close look at the anatomy and behavior of some interesting organisms that live around you unnoticed. Microscopes work like magnifying glasses to give you a closer look at these small organisms.

Different types of microscopes can be used for different purposes. During this exercise, you will be learning to use a **dissecting microscope** to examine larger objects and a **compound microscope** to view smaller specimens.

## ACTIVITY 1

## LEARNING TO USE THE DISSECTING MICROSCOPE

**Caution!**

Dissecting microscopes may have several parts that are not permanently attached! Always carry the scope upright with one hand under the base and the other hand on the arm.

1. Work in groups of **two students**. Get the following supplies: a **dissecting microscope** and a **penny**.

2. Place the penny on the microscope **stage**. Locate the stage by referring to the diagram in **Figure 2-1.**

   Turn the penny so that the Lincoln Memorial is facing you. **Adjust the light** until the penny is brightly illuminated.

3. Set the **magnification control knob**, located on the top or the side of the **head**, to the **lowest setting**.

4. Turn the **focusing knob**, located on the microscope **arm**, until the head is as close to the stage as possible.

   Look through the **ocular lenses (eyepiece)** at the penny. **Turn the focusing knob** until the image of the penny is sharp and clear (the head will be moving **away** from the stage).

   Is Lincoln sitting in the Lincoln Memorial in your penny? _____

5. **While looking through the eyepiece, gradually turn the magnification control knob**. The change in image size will resemble the zoom action of a camera.

**Exercise 2** ■ *Using the Compound and Dissecting Microscopes* 23

**FIGURE 2-1.** Parts of the Dissecting Microscope

## ACTIVITY 2 — MAKING OBSERVATIONS WITH THE DISSECTING MICROSCOPE

1. Work in groups. Get the following supplies: **a small glass bowl, a pipette, and a blunt metal probe**.

2. Use the pipette to remove a **planaria** from the culture container and place it in your bowl. Locate the worm of your choice and draw it into the tip of the pipette (as shown in **Figure 2-2**). When the planaria feels the water current, it will probably form itself into a small protective ball, and it will be easy to draw it into the pipette.

> **Caution!**
> Don't suck the planaria too far into the pipette. It will attach to the inside of the pipette and you won't be able to get it out.
> If additional water is needed in your glass dish, add pond water. Don't use tap water or distilled water.

**FIGURE 2-2.** Drawing Planaria into Pipette

3. **Observe your worm** under the **dissecting microscope**. Use a medium or low light level.

   **Observe carefully** through the microscope for at least **two minutes**.

   Does the worm swim or crawl? _____

   **Describe** the movement of the worm's **muscles** as it moves forward.

**Exercise 2** ■ *Using the Compound and Dissecting Microscopes*     **25**

4. **Draw a picture** of your worm. Make it **large** and **clear. Label** the **head** and the **eyes**.

| DRAWING OF PLANARIA |
|---|
|  |

5. Planaria are common inhabitants of freshwater ponds and streams. On the basis of your **observations**, how do you think they get their food? **Explain** your answer.

6. Hold the **blunt probe motionless**, directly in the path of a moving worm. How does the worm respond?

7. Touch the head end **gently** with the blunt probe. How does the worm respond?

8. Touch the planaria **gently** on several other body parts. Record your observations.

9. Was the **touch response** on other body parts **similar or different** from the response you observed when the worm was touched on the head end?

 Check your answers with your instructor before you continue.

## ACTIVITY 3 — GETTING FAMILIAR WITH THE COMPOUND MICROSCOPE

1. Work in groups. Get a **compound microscope**.

> **Caution!**
> Always carry the scope upright with one hand under the base and the other hand on the arm.

2. Plug the microscope into an electrical outlet and **turn the light on**. The light switch is located on the **base** of the microscope.

3. The compound microscope consists of a system of **optics** (lenses and mirrors) and focusing controls.

   The base and the arm support a **body tube** that **houses the lenses** that magnify the image.

**Exercise 2** ■ *Using the Compound and Dissecting Microscopes*

4. At the top of the scope is the **ocular lens (eyepiece)**. The ocular is only one of a series of lenses that magnify the image. The ocular lens makes the image **ten times larger** than life size (abbreviated 10×).

   **Look through the ocular lens**. You will see a black pointer. Rotate the eyepiece **while looking through the lens**.

   What happens to the pointer? _____

5. At the bottom of the body tube is a revolving **nosepiece**. The lenses that screw into the nosepiece are called **objective lenses**.

   Turn the nosepiece until you hear one of the objective lenses **click into position** (see **Figure 2-3**).

When an objective lens clicks into position, it is in the proper alignment for light to pass from the light source, through the objective, through the ocular lens, and into the viewer's eye.

Turn the nosepiece again to bring a different objective lens into position.

**FIGURE 2-3.** Revolving Nosepiece and Objective Lenses

6. **How will you know** when you have placed the objective lens in the proper position?

7. Note that each objective lens is of a different length. Each of the lenses has a different magnifying power. The **shortest lens** has the **lowest magnifying power** and the **longest lens** has the **highest**.

   Since light from the specimen passes through **both** an objective lens **and** the ocular lens, the **total magnification** of the image is the result of the objective lens magnification **multiplied** by the ocular lens magnification.

8. Using the information on the previous page, **complete the magnification table**.

| Length of Objective Lens | Magnification of Objective Lens | Magnification of Ocular Lens (Eyepiece) | Total Magnification of Specimen |
|---|---|---|---|
| Short | 4× | | |
| Medium | 10× | | |
| Long | 40× | | |

9. On both sides of the arm, you will see the **focus knobs**. The larger ring is the **coarse focus knob** and the smaller ring is the **fine focus knob**.

   Turn the microscope so that you are looking at it **from the side**. Looking at the lenses and stage, turn the **coarse focus knob** all the way in one direction and then reverse the process. **What happens?**

   **Rapidly** turn the **fine focus knob** all the way in one direction and then reverse the process. **What happens?**

10. The **mechanical stage** is a moveable platform designed to hold a microscope slide.

    The distance between the stage and the objective lens is adjusted by using the coarse and fine focus knobs.

    Two control knobs move the mechanical stage. One moves the stage left and right; the other moves the stage forward and back. **Always move the stage using the control knobs.** Don't attempt to move the slide using your fingers.

11. In the center of the stage, you will see the glass of the **condenser lens**, which focuses the light on the specimen.

12. Directly above the light source, you will see a lever that moves from left to right. This lever is connected to the **iris diaphragm**. Looking at the condenser lens, move the **iris diaphragm lever** all the way to the **left**. **What happens?**

13. Looking at the side of the microscope, move the **iris diaphragm lever** all the way to the **right**. **What happens** now?

**Exercise 2** ■ *Using the Compound and Dissecting Microscopes* 29

# ACTIVITY 4            PARTS OF THE COMPOUND MICROSCOPE

Now that you have some experience with the parts of the microscope and their functions, **label the parts** of the microscope on **Figure 2-4**.

**FIGURE 2-4.** Parts of the Compound Microscope

# ACTIVITY 5 — LEARNING TO USE THE COMPOUND MICROSCOPE

1. Work in groups. Get the following supplies: a **prepared slide** labeled **letter "e"** and a piece of **lens paper**.

2. **Clean the letter "e" slide** with the lens paper.

   Using the lens paper, **clean the ocular lens**.

   Be sure that the letter "e" is **right-side up**, in its **normal reading position**, before you place it on the stage.

3. Move the **iris diaphragm** lever to about the middle position. Click the **scanning (lowest power) objective lens** into place.

4. Turn the **coarse focus** until the objective lens is as close to the slide as it will go.

   **While looking through the eyepiece, slowly** raise the objective lens to bring the letter "e" into focus.

## ✓ Comprehension Check

1. If the letter "e" is not lighted brightly enough for you, **what should you do?**

2. If the light is too bright, causing glare and eyestrain, **what should you do?**

3. What is the **total magnification** of the letter "e" image? **Explain** your answer.

**Check your answers with your instructor before you continue.**

**Exercise 2** ■ *Using the Compound and Dissecting Microscopes*

4. Draw the letter "e" **exactly** as it appears under low power. Make it **large** and **clear**.

| LETTER "e" |
|---|
|  |

5. How is the **orientation** of the letter "e" as seen through the microscope **different** from the way an "e" **normally** appears? List **two** differences.

6. **While looking through the eyepiece**, move the stage to the **left**.

   What direction does the image appear to move? _____

   **While looking through the eyepiece**, move the stage **away from you**.

   What direction does the image appear to move? _____

If you wanted to **center** the letter "e" in this drawing, which **two directions** would you have to move the stage?

_____ and _____

7. **Centering the specimen is** *absolutely necessary* **before you can change to a higher power objective!** Do you know why?

   The reason is simple. The more powerful the magnifying lens, the smaller area you see, but you see that small area in greater detail. The area you can see at one time is called the **field of view**. So you could say that the **higher** the power of the objective, the **smaller** the field of view.

8. Center the letter "e" image, rotate the nosepiece, and **click in the 10× objective lens** (also called the **low power objective**).

> **Note:**
> Do not move the coarse adjustment knob!

9. Looking through the eyepiece, can you still see the letter "e"? _____

   Is the letter "e" exactly in the center of the field of view? _____

   If not, **move the slide slightly** to center the image.

   Is the letter "e" sharp and clear? _____

   If not, **gradually adjust the fine focus knob** until the problem is corrected.

   Do you have enough light? _____

   If not, **gradually adjust the iris diaphragm**.

   How has the **image of the letter "e" changed** from the way it looked using the 4× objective lens?

10. Repeat the instructions in **Step 9**, this time changing **from 10×** to the **high-power objective** (40×).

    Looking through the eyepiece, can you see the entire letter "e"? _____

**Exercise 2** ■ *Using the Compound and Dissecting Microscopes*    33

11. **(Circle one answer.)** Your field of view on high power is **larger/smaller** than the field of view on 10×.

    How does the size of the field of view determine how much of the letter "e" you can see?

12. When using the **high-power objective** (40×), what is the **total magnification** of the letter "e" image? _____ **Show your work and explain your answer.**

 **Check your answers with your instructor before you continue.**

## ACTIVITY 6 — PREPARING TEMPORARY SLIDES—WET MOUNTS

### Preparation

A wet mount is a method of preparing a slide that will be used only for a short time. Unlike the letter "e," which was permanently attached to the slide, a wet mount is made by placing the specimen into a drop of liquid on a slide. The specimen and water droplet are held in place by a coverslip.

### Human Epithelial Cells (Cells from the Inside of Your Cheek):

1. Work in groups. Your group will make **two different wet mounts** of cheek cells.

   One will be made with a drop of **stain**. The second will be made by substituting a drop of **physiological saline** for the stain.

2. Get the following supplies: a **dropper bottle of iodine stain or physiological saline, a slide, a coverslip**, and a **clean toothpick**.

3. **Place a drop of iodine stain or a drop of saline near the center of a clean slide.**

4. **Gently** scrape the inside of your cheek with the **end of a toothpick**.

> ### Caution!
> If you scrape too hard, you will be examining blood cells instead of epithelial cells!

You have removed some of the cells that form a **protective covering** for the inside of the mouth. Like other epithelial cells, these are constantly being worn off and replaced by new cells of the same type.

5. **Spread the material from the toothpick** into the drop of water or stain. Add a coverslip.

   **Put two microscopes together** (yours and your partner's) on the laboratory counter so that you can view and compare the stained and unstained slides.

> ### Hint:
> Unstained cells are clear. They are only visible with very low light levels. If you think there are no cells on your slide, adjust the iris diaphragm!

6. **View both slides. Begin with the 4× objective.** Continue until you have located the cells using **all three objective lenses**.

7. **Draw a picture** of **one** cheek cell, **viewed on high power**. Make it **large and clear**.

| HUMAN CHEEK CELL |
|---|
|  |

**Exercise 2** ■ *Using the Compound and Dissecting Microscopes*      **35**

8. Label the following cell structures (referred to as **organelles**) in your drawing of the cheek cells:

| Organelles to Label | Function |
|---|---|
| Nucleus | directs all cell activities |
| Cell membrane | controls movement of materials in and out of the cell |
| Cytoplasm | jelly-like fluid found between the nucleus and cell membrane |

9. Give an example of a **medical procedure** in which epithelial cells are scraped from another area of the body and examined microscopically.

10. Is there an advantage in **using a stain** to view cells microscopically? _____ If so, how is the stain helpful?

## Daphnia—An Aquatic Organism

The water flea, daphnia, is a microscopic organism commonly found in ponds, lakes, and streams. Because it is small and transparent, living daphnia can be studied easily in the laboratory. Daphnia feed on microscopic food particles. Their five pairs of legs are modified into strainers that filter the food particles from the water. Daphnia, in turn, are an important part of the food chain. Many fishes and even larger aquatic animals feed on daphnia.

1. Work in groups. Get the following supplies: a **depression slide** and a **bottle of methyl cellulose** (the bottle may also be labeled **Protoslo®**).

2. Put a **small** drop of the methyl cellulose in the depression on your slide.

    Use the pipette in the culture jar to remove a **daphnia** from the container and place it in the depression on the slide. **You don't need a coverslip.**

3. Make your observations using the **scanning lens** of the microscope (4×).

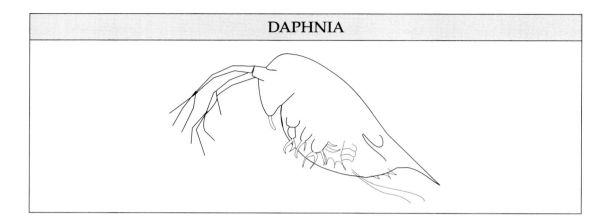

4. Add the following details to the daphnia outline below: **eye, heart, brain, and intestine. Label the details you put in the drawing. Also, label the head and legs**.

5. To see details of their internal structure, both the daphnia and the cheek cell must be viewed through the microscope. Since they are both so small, why were you able to see organs such as the brain and heart in the daphnia, but not in the cheek cell? **Explain** your answer.

> ### Hint:
> **Don't forget to make sure your hypothesis is testable!**

**Exercise 2** ■ *Using the Compound and Dissecting Microscopes*    **37**

Self Test

1. Complete the table by entering the appropriate part of the compound microscope or the correct function for the stated microscope part.

| PARTS OF THE COMPOUND MICROSCOPE | FUNCTION |
| --- | --- |
| Ocular lens (eyepiece) | |
| Stage | |
| | Focuses light on the specimen |
| Nosepiece | |
| | Objective lens used to first locate a specimen |
| | Regulates the amount of light that passes through the specimen |
| Fine focus knob | |
| | Objective lens with the lowest magnifying power |
| | Objective lens with the highest magnifying power |
| Coarse focus knob | |

2. What is the **total magnification** if the ocular lens is 15× and the objective lens is 20×? **Show your work!**

3. List **three differences** between the dissecting microscope and the compound microscope:

   a.

   b.

   c.

4. **What type of microscope** (compound or dissecting) would you use to observe the following:

   _____ Cells from the lining of your stomach

   _____ A splinter in your finger

_____ A cockroach you found in the kitchen

_____ Mildew from your shower curtain

5. Suppose you were watching a daphnia under the microscope and noticed that it moved **toward you** and then **to your right**.

   Which direction(s) did the daphnia **actually** move? _____

   **Explain your answer**.

6. Explain how you would **correct the following problems** experienced when using a microscope:

   a. When changing magnification from 10× to 40×, the specimen disappears.

   b. The field of view is too dark.

   c. Your field of view is partially obscured by a dark area.

   d. There is a fingerprint in your field of view.

   e. There are many hollow, dark circles in your field of view.

# EXERCISE 3

# Functions and Properties of Cells

## Objectives

After completing this exercise, you should be able to:
- identify and explain the functions of the major cellular organelles
- explain the similarities and differences between plant and animal cells
- explain the concepts of diffusion and osmosis and why they are important to cell physiology
- use indicator chemicals to test for the movement of molecules and to determine the direction of diffusion
- explain the process of osmosis in living cells exposed to different extracellular solute concentrations
- apply your knowledge of cell structure and function to real-life situations.

## CONTENT FOCUS

Typically, when you look at a plant or animal, it is easy to think of an organism as one large unit. When looking through the microscope, however, it becomes obvious that an organism is composed of trillions of tiny units called **cells**, which work cooperatively to carry out the functions that keep us alive. If the activity of your cells stopped, even for a few moments, death could quickly follow.

There are about 200 different types of cells in the human body. These cells can be quite different in shape, structure, and function, but they all have some basic characteristics in common.

Within each cell lies a collection of specialized structures called **organelles**, in which particular chemical activities take place. The various cell structures carry out activities that mirror the functions of our body organs. Most organelles are compartments bounded by membranes.

Learning about cell organelles will be more meaningful when you understand the role that each part plays in the life of a cell. **Table 3-1** gives a brief summary of the functions of some basic organelles.

### TABLE 3-1
### SOME CELL ORGANELLES AND THEIR FUNCTIONS

| ORGANELLE VISIBLE WITH THE LIGHT MICROSCOPE | FUNCTION |
|---|---|
| Cell wall | External support and protection; if present, located outside the cell membrane |
| Cell membrane | Surrounds the cytoplasm; barrier between the environment and the cell that controls the movement of materials in and out of the cell |
| Cytoplasm | Liquid/gel "filler" substance inside the cell membrane; all additional cell organelles are suspended in the cytoplasm |
| Nucleus | Stores and transfers information (DNA and RNA) needed to control cells functions |
| Chloroplast | Structure in which photosynthesis takes place; contains chlorophyll and is green in color |
| **ORGANELLE NOT VISIBLE WITH THE LIGHT MICROSCOPE** | **FUNCTION** |
| Mitochondria | Location of aerobic cellular respiration; produces ATP energy |
| Ribosomes | Protein synthesis; ribosomes may be found free in the cytoplasm or attached to the endoplasmic reticulum (ER) |
| Endoplasmic reticulum (ER) | Membrane manufacturing complex |
| Rough endoplasmic reticulum (ER) | Ribosomes attached to ER membrane; protein synthesis |
| Golgi apparatus | Packaging of molecules for export out of the cell; most exported molecules are proteins |

## ACTIVITY 1 — OBSERVING MOVEMENT INSIDE LIVING CELLS

1. Work in groups. Get **two compound microscopes** and set them up, side by side.

2. Make **two wet mounts**, one of **amoeba** and one of **elodea**. Place one wet mount on each microscope.

    **Amoeba** is a **one-celled organism** commonly found in pond water. **Elodea** is a **freshwater plant** frequently used in home aquariums.

# Exercise 3 ■ *Functions and Properties of Cells*

3. **Make a drawing** of each organism, **labeling** all the cell parts you can identify.

   Use the information in **Table 3-1** to help you locate and identify the cellular organelles.

   | AMOEBA | ELODEA |
   |--------|--------|
   |        |        |

4. Which organism(s) can carry on photosynthesis? _____

   What **evidence** did you see that led you to draw this **conclusion?**

5. Which organism(s) has a **cell wall?** _____

6. In which organism(s) are you able to observe the cytoplasm moving? _____

7. Some organisms can secure food by surrounding their prey with cell extensions. This process is called **phagocytosis** (shown in **Figure 3-1**).

   Which organism(s) do you think would be able to do this? _____

   **Explain your answer.**

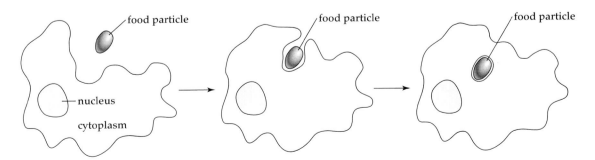

**FIGURE 3-1.** Phagocytosis

8. Which **cell organelles** can be seen changing position in the moving cytoplasm of each organism?

   elodea _____

   amoeba _____

9. Which organism demonstrates **locomotion** (ability to move from place to place?
   _____

   Write a **clear description** of the **movement pattern** of this organism.




■ **Check your answers with your instructor before you continue.**

## ACTIVITY 2  WATER MOLECULES IN MOTION

*Preparation*

You have just observed the movement of cytoplasm inside cells. Everywhere around us, molecules of liquids and gases are in constant, random movement.

Of course, these moving molecules are much too small to be observed directly, but we can demonstrate this activity by placing some colored powder into a drop of water.

1. Work in groups. Get the following supplies: **a slide, a coverslip, a dropper bottle of distilled water, a toothpick, and carmine powder**.

2. Place a drop of distilled water on the slide. **Carefully** add a tiny amount of carmine powder to the water drop. Use the **pointed tip** of the toothpick to scoop up the carmine powder.

   Imagine how much carmine powder you think you might need, and then scoop up half that amount.

   Add a coverslip and observe under low power, then medium power, and finally, under **high power**.

**Exercise 3** ■ *Functions and Properties of Cells*       **43**

3. **Describe** the activity of the carmine particles in the water droplet.

4. Since carmine particles are **not** alive, what is **causing** the carmine particles to move?

   As was mentioned earlier, molecules of liquid are always in motion. This means water molecules everywhere are constantly moving and bumping into each other.

   As they move, they will also collide with anything floating in the water, such as the carmine particles.

5. **(Circle one answer.)** If we **warmed up** the carmine slide a little bit, the rate of molecular motion would be **faster/slower/stay the same**.

   **Explain your answer.**

   You have just observed that molecules are in constant motion. Movement of molecules will allow us to explain **diffusion and osmosis**, two important cell processes.

   The process of **diffusion** occurs whenever dissolved particles move from an area of **high concentration** (more of them) to nearby areas where they are **less concentrated**. The diffusion process is similar to a bumper car ride (see **Figure 3-2**). Cars are moving quickly and often collide. The force of the collision sends the cars shooting outward into an empty area of the bumper car rink. The more an area is crowded with bumper cars, the more collisions occur.

   In this way, the cars are gradually dispersed from the crowded center of the rink (an area of **high** bumper car concentration) to the emptier fringe areas (areas of **lower** bumper car concentration). This is a good model of what happens when particles diffuse in the environment (see **Figure 3-2**).

   **Diffusing molecules always move outward from an area of high concentration into areas of lower concentration.**

   A similar process to diffusion occurs with **water** molecules in the environment. Cells are surrounded by a membrane that allows some substances to pass through but not others. This is referred to as a **selectively permeable membrane**.

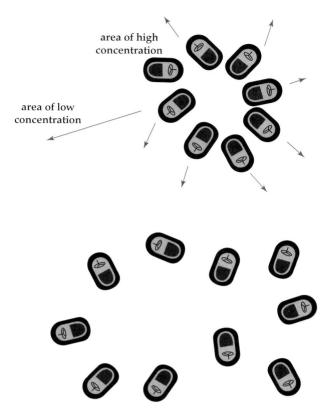

**FIGURE 3-2.** Bumper Car Model of Diffusion

Water molecules disperse through a **selectively permeable membrane** from an area of **high concentration** to an area of **lower concentration**. This process is given a different name, **osmosis**. We can demonstrate the processes of diffusion and osmosis in an experiment using **dialysis tubing**, an artificial membrane with many small pores.

## ACTIVITY 3                                        DIFFUSION AND OSMOSIS

1. Work in groups. Get the following supplies: one plastic **jar, one** piece of **dialysis tubing, and two long pieces of thread**.

   Dialysis tubing is an **example** of a selectively permeable membrane. **Cell membranes** are also selectively permeable.

2. **Wet the piece of dialysis tubing** for a minute or two and open it as demonstrated by your instructor.

   Referring to **Figure 3-3**, construct a **dialysis tubing bag** by tying one end **tightly** with the thread. Wrap the thread several times around the tubing before tying to prevent leakage.

# Exercise 3 ■ *Functions and Properties of Cells*

**FIGURE 3-3.** Dialysis Tubing Bag

3. Get a bottle of **stock solution.** The solution consists of **water** with a mixture of dissolved **salt** (NaCl), **sugar** (glucose), and **starch.**

> **Hint:**
> Starch tends to settle out on the bottom of the bottle. Make sure you shake the stock solution to mix completely before measuring.

Using a graduated cylinder, measure 20 **milliliters (ml)** of stock solution and pour it into your dialysis tubing bag.

Using the remaining piece of string, tightly **tie the open end** of the dialysis tubing.

4. **Weigh your bag** according to the following directions, using the balance in your classroom:

   a. Pat the bag dry. Cut off any excess string.
   b. Make sure the **balance is set to zero.** (If you have any problems using the scale, consult your laboratory instructor.)
   c. Weigh the bag to **the nearest tenth of a gram.**

   Record the **initial weight** here: _____

5. Fill the plastic jar about three quarters full with **distilled water.** Do not fill the jar with tap water. **Immerse** the dialysis tubing bag in the jar of distilled water.

6. **Write a hypothesis.** What do you think will happen to the weight of the bag during the 20 minutes it sits in the jar of distilled water?

7. Set the jar aside for **at least 20 minutes.**

   **While you are waiting, complete the Comprehension Check and then set up the controls for chemical testing in Activity 4.**

   You will return to your dialysis experiment when the 20 minutes have elapsed.

## ✓ Comprehension Check

1. When someone uses a 10% glucose solution, what is actually in this solution? Fill in the blanks with the correct percentages (this must add up to 100%).

   _____ % glucose _____ % water

2. **Figure 3-4** represents a dialysis tubing bag containing dissolved salt, sugar, and starch, surrounded by distilled water.

**FIGURE 3-4.** Dialysis Bag Surrounded by Distilled Water

(**Circle one answer.**)

The **highest** concentration of **salt** is inside the bag/outside the bag.

The **highest** concentration of **sugar** is inside the bag/outside the bag.

**Exercise 3** ■ *Functions and Properties of Cells* 47

The **highest** concentration of **starch** is inside the bag/outside the bag.

The **highest** concentration of **water** is inside the bag/outside the bag.

 **Check your answers with your instructor before you continue.**

## ACTIVITY 4 — CONTROLS FOR THE DIFFUSION AND OSMOSIS EXPERIMENT

### Preparation

You will perform **three** different chemical tests to determine the results of your experiment. Each uses an **indicator chemical that changes color** if starch, salt, or sugar is present.

a. **Iodine Test for the presence of starch.** When added to a solution, iodine will turn **black** if starch is present. If no starch is present, the iodine will remain **reddish-brown** in color.

b. **Silver Nitrate Test for the presence of chloride ions.** When added to a solution that contains chloride ions, the silver nitrate will change color from **clear** to **cloudy white**. (Remember: salt is composed of sodium ions and chloride ions.)

c. **Benedict's Test for simple sugars.** When added to a solution that contains simple sugars, such as glucose, Benedict's solution will change color from **turquoise blue** to one of the following colors: **green, yellow, orange, or red**. Green represents the smallest amount of simple sugar and red the highest.

> **Note:**
> It is important to remember that the starch, salt, and glucose are not changing in any way. It is the indicator chemicals that are changing color.

1. Work in groups. Get a **large beaker.** Fill it **one-third full** with **tap water** and heat it to boiling.

   While you are waiting for the water to boil, set up the materials needed for chemical testing.

2. To set up your **control** tubes, get the following supplies: a **test tube rack, a ruler, a marking pencil, a test tube holder, and three test tubes**.

   **Label the test tubes salt, sugar, and starch.**

3. Using the ruler, place a **line** on each test tube **1 cm from the bottom** of the tube.

   **Fill each of the three test tubes** to the 1-cm line with stock solution.

4. To the test tube labeled **salt**, add **one dropper full** of **silver nitrate** solution.

## Caution:
**Avoid getting this solution on your skin.**

5. Shake the tube gently to mix the contents. **Observe the color** of the solution. **Record your results** in the appropriate columns in **Table 3-2**.

### TABLE 3-2
### RESULTS OF CHEMICAL TESTS

| INDICATOR SOLUTION | SHOWS THE PRESENCE OF | INITIAL COLOR | FINAL COLOR | RESULTS (+ OR −) |
|---|---|---|---|---|
| Silver Nitrate Test Control Tube | | | | |
| Iodine Test Control Tube | | | | |
| Benedict's Test Control Tube | | | | |
| Silver Nitrate Test Experimental Tube | | | | |
| Iodine Test Experimental Tube | | | | |
| Benedict's Test Experimental Tube | | | | |

Exercise 3 ■ *Functions and Properties of Cells*          49

> **Note:**
> Save all three of your control test tubes (salt, starch, and sugar) for later use.

6. To the test tube labeled starch, add **one dropper full** of **iodine** solution. Shake the tube gently to mix the contents.

   **Observe the color** of the solution. **Record** your results in **Table 3-2**.

7. To the test tube labeled sugar, add **one dropper full** of **Benedict's** solution. Shake the tube gently to mix the contents.

8. **Carefully** lower the test tube into the boiling water bath and allow it to remain for **two to five minutes**.

> **Caution:**
> Hot glass looks exactly like cold glass. Use a test tube holder to remove test tubes from the boiling water. Don't leave the test tube holder clamped on the test tube while in the boiling water. Hot metal looks like cold metal!

9. **Using a test tube holder**, remove the tube from the boiling water bath and place it in the test tube rack.

10. **Observe the color** of the solution. **Record** your results in **Table 3-2**.

## ACTIVITY 5 — COMPLETING THE DIFFUSION AND OSMOSIS EXPERIMENT

1. **Remove** your dialysis tubing bag from the jar of water. To test for **osmosis**, follow the **same weighing directions** as in **Activity 3, Step 4**, and weigh your bag again.

2. Record the **final weight** here: _____

   Did the bag **gain or lose weight?** _____

   Was your hypothesis about the **weight change** correct? _____

3. Calculate the **percent change in weight** and record your answer.

   $$\text{percent change} = \frac{\text{final weight} - \text{initial weight}}{\text{initial weight}} \times 100$$

   percent weight change = _____ %

   What do you think **moved through the membrane** to cause this result?

4. To test for **diffusion**, get **three more test tubes. Label the test tubes salt, sugar, and starch**.

   Using the ruler, place a **line** on each test tube **1 cm from the bottom** of the tube.

   **Fill** each of the three test tubes to the **1-cm line** with **water from the plastic jar that held the dialysis tubing**.

   **Test the water** for salt, starch, and sugar using the same indicator chemicals you used when setting up the controls in **Activity 4**. Each test tube will contain **only** water from the jar and the indicator chemical.

   **Do not add stock solution to these test tubes!**

5. **Record** the results of all three tests in **Table 3-2**.

   According to your test results, **list the molecules** that were able to **diffuse** out of the dialysis bag.

6. If molecules are **expected** to move from **high concentration** to **low concentration**, why didn't **all** the molecules (salt, sugar, and starch) leave the dialysis tubing bag?

7. The ability to separate molecules from each other using a **selectively permeable** membrane is referred to as **dialysis**. Did the experiment you just finished illustrate this concept? **Explain your answer.**

   Check your answers with your instructor before you continue.

**Exercise 3** ■ *Functions and Properties of Cells*

## ACTIVITY 6 — THE RELATIONSHIP BETWEEN SOLUTE CONCENTRATION AND OSMOSIS IN RED ONION CELLS

*Preparation*

As you have seen, a solution has two components: dissolved molecules (called **solutes**) and water (the **solvent**). The cell membrane, which controls the movement of materials in and out of the cell, is not able to control the movement of all molecules. Exceptions include carbon dioxide, oxygen, and water. Water, therefore, can move passively through the cell membrane following the concentration gradient.

In your previous experiment with the dialysis tubing, you observed that water will move through a cell membrane from an area of high concentration to an area of lower concentration. The solution with a **higher solute concentration** (in our example, this was the solution **inside** the dialysis tubing bag) is said to be **hypertonic** to the solution with a lower solute concentration (the water in the beaker). You can reverse this statement and say that the water in the beaker is **hypotonic** (has a **lower solute concentration**) to the fluid inside the dialysis tubing bag. Solutions with **equal solute concentrations** are said to be **isotonic** to each other.

In the following experiment, you will observe the movement of water across the cell membrane of a red onion cell. In red onion cells, the sap vacuole contains a purple pigment that will make it easier for you to observe osmosis.

1. Work in groups. You will be making **two different wet mounts** of onion cells.

    One will be made with a drop of **distilled water**. The other will be made by substituting a drop of **20% saline** for the water.

    Get the following supplies: **a piece of onion bulb, a scalpel, forceps, a dropper bottle of distilled water, a dropper bottle of 20% saline solution, two slides, and two coverslips**.

2. Make wet mounts of the onion-skin cells, following the directions in **Figure 3-5**.

    With a marking pencil, **write** "S" on the right side of the slide to indicate saline solution.

Pull off a section of a **single leaf** from the **purple** skin on the **outer surface** of the onion bulb.

With forceps remove a piece of that thin **purple** skin and place it in the drop of **water** or **salt solution** on your slide.

The piece of onion skin you remove **must be thin and transparent**. It will look like a small piece of purple plastic wrap.

Gently lower a coverslip over the specimen.

**FIGURE 3-5.** Wet Mount of Onion Skin

**Exercise 3** ■ *Functions and Properties of Cells* **53**

3. Place **two microscopes** side by side and **compare** the two slides. **Start with the 4× scanning lens.** Center and focus the specimen. Change to **low**, and then to **high** power.

4. Wait **five minutes**, then check on both slides again.

   **Draw a large picture** showing **one or two representative cells** from each slide.

   On each diagram, **draw arrows** showing the **direction of water movement** through osmosis.

| ONION IN DISTILLED WATER | ONION IN 20% SALINE SOLUTION |
|---|---|
|  |  |

5. Add the following labels to your diagrams: **cell wall, nucleus, nucleolus, cytoplasm**.

   Identify and label each solution that the onion cells were exposed to on the slide as **hypertonic, hypotonic, or isotonic**.

6. In one of your slides, you should be able to see the onion **cell membrane**.

   On which slide is the cell membrane visible? _____

   **Add a label** for the cell membrane to your drawing.

7. **What happened** to the onion cells that were soaked in the **20% saline solution**? **Support your answer** by mentioning details you observed under the microscope.




Check your diagram labels and answers with your instructor before you continue.


Comprehension Check

1. Assume that the interior of an onion cell is **98% water molecules**. The remaining **2%** is composed of various dissolved particles.

    The salt solution you applied to the slide is _____ % **water** molecules and _____ % **salt** particles.

2. **(Circle one answer.)** The process that occurred to the onion cells in salt solution was an example of **diffusion/osmosis**.

    **Explain your answer.**

**Exercise 3** ■ *Functions and Properties of Cells* 55

3. The onion cell membranes allowed water to enter and leave the cells freely, but **did not allow** salt particles to diffuse into the cells.

   This is an example of a _____ **permeable membrane**.

4. A student performed the onion cell experiment using **human blood cells**. When the student placed the blood cells in **salt water**, they shriveled up and died. When the student placed other blood cells in **distilled water**, they filled up, exploded, and died.

   **(Circle one answer.)** The salt water represents a **hypertonic/isotonic/hypotonic** solution compared to the red blood cells.

   **(Circle one answer.)** The distilled water represents a **hypertonic/isotonic/hypotonic** solution compared to the red blood cells.

   What cellular organelle present in onion cells, but **not** blood cells, prevented the leaf cells from shriveling or exploding during your experiment? _____

5. Eye drops or nose sprays are made to be isotonic to your body fluids. What would happen if the solutions were hypertonic to your body fluids? **Explain** your answer.

Self Test

1. Of the cell organelles studied in this exercise, **list three** that are found in plant cells but not in animal cells.

2. In **Figure 3-6**, **draw arrows** showing the movement of the water molecules.

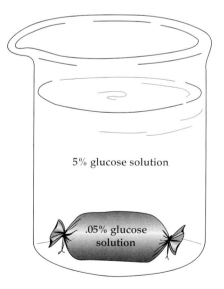

**FIGURE 3-6.** Movement of Water Molecules

**Exercise 3** ■ *Functions and Properties of Cells*     **57**

3. A patient with partial kidney failure enters the hospital. The doctor tells the patient that the kidney is responsible for filtering and removing waste molecules from body fluids. The doctor recommends use of a **kidney dialysis** machine. Using your knowledge of **diffusion, osmosis, and dialysis**, what do you think the dialysis machine will do for this patient?

4. You are admitted to the hospital with severe dehydration. A student nurse is directed to give you a transfusion of **blood plasma** to replace your lost body fluids. The nurse gives you a transfusion of **distilled water** by mistake. **Explain** what will happen to your **red blood cells**. In your explanation, use the following words: **cell membrane, osmosis, high concentration, and low concentration**.

5. Will a sugar cube dissolve faster in iced tea or hot tea? _____

    **Explain your answer**. In your explanation, use the following words: **diffusion, high concentration, low concentration, molecular movement, and collisions**.

6. You are in a restaurant that is divided into "smoking" and "nonsmoking" sections. There is a partition between the two sections, but the partition does not reach the ceiling. Despite the fact that you are seated in the nonsmoking section, you are bothered by the smell of cigarette smoke during your meal. Explain why the cigarette smoke was present in the nonsmoking section of the restaurant. In your explanation, use the following terms: **diffusion, molecules, higher concentration, and lower concentration**.

7. Roads are sometimes salted to melt ice. What effect does this have on plants growing along the road? **Explain** your answer.

EXERCISE 4

# Investigating Cellular Respiration

## Objectives
After completing this exercise, you should be able to:
- explain the process of aerobic cellular respiration in plants and animals
- summarize experimental results that show that microorganisms also perform aerobic cellular respiration
- compare the rate of aerobic cellular respiration and carbon dioxide production while at rest with that during exercise
- explain the similarities and differences between aerobic respiration and anaerobic respiration and give examples
- explain the similarities and differences between ethanol fermentation and lactic acid fermentation
- apply your knowledge of aerobic and anaerobic respiration to real-life situations.

## CONTENT FOCUS

Cellular respiration is the process by which living organisms convert the **chemical energy** in food (**organic molecules**) to a useful form for cellular functions. This conversion process is comparable to the production of electrical energy to run machines in your home. A television cannot take energy directly from coal, gasoline, or even nuclear fuel. These fuels must be converted to electricity by your community power plant before they can be used by your home appliances.

Cells must transfer the energy stored in food molecules into **adenosine triphosphate (ATP)**, a form of energy that can be used to run cell activities. If you read the ingredients on a candy wrapper or a box of corn flakes, however, you will not find ATP listed. The energy in the foods has to be **transferred** into ATP. This conversion process is called **aerobic cellular respiration**.

During **aerobic respiration**, living organisms extract energy from the chemical bonds in food molecules (such as carbohydrates, fats, or proteins) and convert that energy into ATP.

The process of cellular respiration can be summarized with this equation:

food molecules + oxygen ⟶ carbon dioxide + **ATP energy** + water + heat

Aerobic cellular respiration is not the same as breathing. It occurs continuously, day and night, in all living cells. If it stops, the cell will quickly die. Aerobic respiration is not a single chemical reaction, but involves as many as 50 intermediate steps and the formation of a number of different compounds before carbon dioxide and water are finally produced. Specific enzymes are needed to control each step. For all living organisms, the chemical reactions of cell respiration are amazingly similar. Many pathways are virtually identical, even between such dissimilar organisms as bacteria and human beings.

The process of aerobic cellular respiration **uses oxygen** and **produces carbon dioxide**. For this reason, cellular respiration cannot take place without gas exchange. Thus, when you breathe in, oxygen is carried by the bloodstream to all the cells of your body. As blood circulates through your tissues, it picks up the carbon dioxide produced by cellular respiration and transports it to the lungs to be removed. Every time you exhale, carbon dioxide is released.

To sum up, aerobic respiration is a process that uses oxygen to burn the food (fuel) we consume and produce useful ATP energy. The more energy your body needs, the more fuel and oxygen you must take in. It should now be clear why you increase your breathing rate (and therefore your oxygen intake) when you exercise. During exercise, you also increase the amount of carbon dioxide you produce, since you burn more glucose to produce ATP.

## ACTIVITY 1 — EVIDENCE THAT $CO_2$ IS RELEASED DURING CELLULAR RESPIRATION

*Preparation*

From the following equation, you can see that organisms produce carbon dioxide gas ($CO_2$) during aerobic cellular respiration.

food molecules + oxygen ⟶ **carbon dioxide** + ATP energy + water + heat

**Bromothymol blue** is a solution that turns yellow when $CO_2$ is added and turns back to blue when $CO_2$ is removed. We will use bromothymol blue as an **indicator** that will tell us if $CO_2$ is produced.

**Exercise 4** ■ *Investigating Cellular Respiration*

1. Work in groups.

   In your classroom, you will find a demonstration of cellular respiration. The demonstration setup consists of a paper bag covering a rack of sealed test tubes containing **bromothymol blue and sample organisms (see Figure 4-1)**.

   > **Note:**
   > Read through the instructions COMPLETELY and record your hypotheses in Table 4-1 BEFORE LOOKING AT THE DEMONSTRATION!!

   Demonstration Setup:

   | | |
   |---|---|
   | Tube 1 | No sample |
   | Tube 2 | Seedling (young plant) |
   | Tube 3 | Plant seeds |
   | Tube 4 | Plant leaf |
   | Tube 5 | Cricket |
   | Tube 6 | Pebble |

   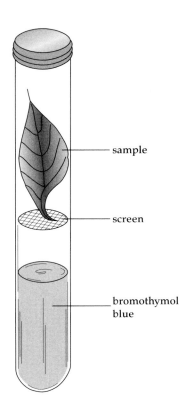

   The samples are suspended on a piece of screen so that they will not fall into the bromothymol blue solution.

   **FIGURE 4-1.** Appearance of Experimental Test Tubes

2. **Form a hypothesis for each test tube.** Which of the samples will produce $CO_2$ and change the color of the indicator solution from blue to yellow?

   Enter your hypotheses in the first column of **Table 4-1**.

| TABLE 4-1 WHICH TUBES CONTAIN CARBON DIOXIDE? | | | |
|---|---|---|---|
| Sample | Hypothesis Will $CO_2$ Be Produced? (Yes or No) | Results Color of Indicator Solution | Conclusion Was $CO_2$ Produced? (Yes or No) |
| Tube 1: No sample | | | |
| Tube 2: Seedling (young plant) | | | |
| Tube 3: Plant seeds | | | |
| Tube 4: Plant leaf | | | |
| Tube 5: Cricket | | | |
| Tube 6: Pebble | | | |

3. Carefully lift the paper bag and **observe** the samples. **Record your observations** in the **Results column** of Table 4-1.

## Comprehension Check

1. On the basis of your observations, which tubes showed evidence of $CO_2$ production? Record your answers in the **Conclusions column of Table 4-1**.

2. Was there a control tube in this experiment? _____

    If so, which tube was the control? _____ **Explain your answer**.

3. On the basis of your results, which of the sample organisms was performing cellular respiration? **Explain your answer.**

4. **(Circle one answer.)** If one test tube contained a **dead cricket**, the amount of $CO_2$ in the test tube would be **more/less/exactly** the same compared to the tube with the living cricket.

5. **(Circle one answer.)** Seeds contain plant embryos. On the basis of the results of this experiment, the embryos in the tested seeds are **active/dormant (inactive)**.

 **Check your answers with your instructor before you continue.**

## ACTIVITY 2 — DO MICROORGANISMS PERFORM CELLULAR RESPIRATION?

Preparation

Yeasts are microscopic fungi that are commercially very important. They perform aerobic cellular respiration using pathways similar to those found in larger plants and animals.

**Methylene blue dye** can be used as an **indicator** for cellular respiration in yeast.

Aerobic respiration releases hydrogen ions and electrons that are picked up by the methylene blue dye, gradually turning the dye **colorless**. The mitochondria of yeast cells undergoing cell respiration will appear as a clear area surrounded by a ring of light blue cytoplasm. The nucleus may be visible as a small, darkly stained spot.

If cellular respiration is not taking place, the mitochondria will absorb the blue dye and will not turn colorless. The cells will appear to have a large, darkly stained central area surrounded by the ring of light blue cytoplasm.

1. Work in groups. Get the following supplies: **slides, coverslips, yeast suspension, a pipette, and a dropper bottle of methylene blue dye**.

2. Place a **drop of yeast suspension** on a clean microscope slide. Add **one small** drop of **methylene blue** dye and place a **coverslip** over the mixture. Observe the yeast cells with the **high-power objective**.

3. Is cell respiration occurring in any yeast cells on your slide? _____

   What percentage of the yeast cells is **not** undergoing cell respiration? _____ (Make an estimate.)

   Why **isn't** cell respiration taking place in all cells?

4. **Draw** one yeast cell that is undergoing cell respiration and one that **isn't**. Make the drawings **large and clear**. Label the **cytoplasm, nucleus (if visible), and mitochondria** of your yeast cells.

| YEAST CELL CELL RESPIRATION OCCURRING | YEAST CELL CELL RESPIRATION ABSENT |
|---|---|
| | |

**Check your drawings with your instructor before you continue.**

Exercise 4 ■ *Investigating Cellular Respiration*

## ACTIVITY 3 — COMPARISON OF CLASSROOM AIR WITH EXHALED AIR

Preparation

This experiment uses **limewater** (a very concentrated **calcium carbonate solution**) as an **indicator** solution. Limewater turns cloudy when $CO_2$ is added to it.

1. Work in groups. Get the following supplies: **two small beakers, two drinking straws, a rubber bulb, and a container of limewater.**

2. **Fill** each beaker **about half full with limewater.** Label the beakers 1 and 2.

> **Note:**
> Read through the instructions COMPLETELY before continuing.

3. Do you think bubbling **room air into Beaker 1** will cause the limewater to turn cloudy? **Enter your hypothesis in Table 4-2.**

4. Do you think bubbling **your exhaled air into Beaker 2** will cause the limewater to turn cloudy? **Enter your hypothesis in Table 4-2.**

### TABLE 4-2
### EFFECT OF BUBBLED AIR ON LIMEWATER

|  | Hypothesis (Yes or No) | Results |
|---|---|---|
| Will room air cause the lime water in Beaker 1 to turn cloudy? |  |  |
| Will exhaled air cause the lime water in Beaker 2 to turn cloudy? |  |  |

5. Attach the rubber bulb to one of the drinking straws. Using the drinking straw with the bulb, continuously bubble air into Beaker 1 for one **minute.**

   **Observe** the beaker during the bubbling process and **record** your results in **Table 4-2.**

6. Using another drinking straw, **very gently** bubble your exhaled air into **Beaker 2** for **one minute**.

| *Caution!* |
|---|
| **Be careful not to suck the solution into your mouth!** |

**Observe** the beaker during the bubbling process and record your results in **Table 4-2**.

7. When you have completed your experiment, empty your beakers into the **waste container**.

### ✓ Comprehension Check

1. In one or two sentences, **summarize the results** of this experiment.

2. On the basis of your results, **what gas was bubbled** into Beaker 2? _____

3. In reference to **Question 2**, what process **produced** this gas?

**Check your answers with your instructor before you continue.**

**Exercise 4** ■ *Investigating Cellular Respiration*

## ACTIVITY 4 — EFFECT OF EXERCISE ON CARBON DIOXIDE PRODUCTION

### Preparation

Breathing is controlled by a "breathing center" in the medulla of the brain. This center is activated by a **buildup of carbon dioxide in the blood and/or a low oxygen concentration**, such as occurs at high elevations. This experiment will compare the **amount** of $CO_2$ produced during exercise with that produced at rest.

We will use **bromothymol blue** as an **indicator** that will tell us **how much** $CO_2$ is produced. Recall that bromothymol blue turns **yellow** when $CO_2$ is **added** and returns to **blue** when $CO_2$ is **removed**.

1. Work in groups.

   Get the following supplies: **three 400-ml beakers, bromothymol blue solution, ammonia solution, a plastic bag, a coupler tube, a mouthpiece, a rubber band, and an air stone assembly.**

2. Fill each beaker **with 200 ml of bromothymol blue indicator solution. Label** the beakers **1, 2, and 3**.

3. **Attach the coupler tube** to the plastic bag and hold the tube in place with the rubber bands (as shown in **Figure 4-2**).

### Testing Room Air

1. Fill the plastic bag with **room air** by pulling it through the air ("rounding out" the bag).

   **Quickly** plug the bag with the **air stone assembly. Be careful not to lose the air.**

2. Place the **air stone** into the indicator solution in **Beaker 1** (see **Figure 4-2**).

   **Squeeze the air out** of the plastic bag so that it bubbles into the beaker.

**FIGURE 4-2.** Apparatus to Bubble Captured Air into Bromothymol Blue Solution

Ammonia Treatment

1. Using the dropper bottle, add **ammonia** to **the beaker one drop at a time** and **gently** stir after each drop.

   Be sure to **keep an accurate count** of the number of drops added.

2. Continue adding ammonia until the indicator solution changes to a **light green color and remains green for 30 seconds**.

3. **Record** the number of drops you added in **Table 4-3**.

   The amount of ammonia needed to change the indicator solution from **blue to green** will be used as an estimate of **the amount of $CO_2$ present** in the room air.

**Exercise 4** ■ *Investigating Cellular Respiration*

| TABLE 4-3 EFFECT OF EXERCISE ON $CO_2$ PRODUCTION |||
|---|---|---|
| SAMPLE | COLOR AFTER BUBBLING | DROPS OF AMMONIA USED |
| Room air (Beaker 1) | | |
| Exhaled air at rest (Beaker 2) | | |
| Exhaled air after exercise (Beaker 3) | | |

Testing Exhaled Air before Exercise (at Rest)

1. Secure the mouthpiece onto the plastic bag and fill the bag with exhaled air, crimping the end of the bag with your hand to prevent the air from escaping. **Quickly plug the bag with the air stone assembly**, being careful not to lose the air.

2. Place the air stone into the indicator solution in **Beaker 2**. Squeeze the air out of the plastic bag so that it bubbles into the beaker. Keep squeezing until the plastic bag is empty.

   **Add ammonia** one drop at a time until the indicator solution changes to a **light green color and remains green for 30 seconds. Record** the number of drops you added in **Table 4-3**.

Testing Exhaled Air after Exercise

1. Secure the mouthpiece on the plastic bag. Set the bag close at hand and **exercise vigorously for two minutes**. You may jog around the room, do jumping jacks, or step rapidly up and down on an aerobic step.

   **Immediately** fill the plastic bag with exhaled air and plug the bag with the **air stone assembly**.

2. Bubble the exhaled air into **Beaker 3** as you did when testing the exhaled air before exercise. **Add ammonia** until the indicator solution changes to a **light green color and remains green for 30 seconds. Record** the number of drops you added in **Table 4-3**.

## ✓ Comprehension Check

1. What happened to the amount of $CO_2$ in your exhaled air after exercise? **Explain your answer**.

2. What do you think would happen if you exercised for ten minutes instead of only two minutes? **Explain your answer**.

3. When you exercise, your heart beats faster and the blood carries more oxygen to your muscles. What do muscles cells do with this oxygen?

**Check your answers with your instructor before you continue.**

# ACTIVITY 5     ALTERNATIVES TO AEROBIC RESPIRATION

Preparation

You have demonstrated that cells can carry out aerobic respiration when oxygen is available, but they can also produce ATP energy with an anaerobic process called **fermentation** when oxygen is in short supply.

The fermentation pathway can produce a variety of end products (refer to **Figure 4-3**).

In most plant cells and in yeasts, pyruvic acid is converted to ethanol (drinking alcohol) and carbon dioxide. Alcohol fermentation by yeast cells is used commercially to manufacture baked goods and alcoholic beverages.

Alcohol fermentation is summarized with this equation:

food molecules → pyruvic acid → carbon dioxide + ATP energy + *ethanol* + heat

**Exercise 4** ■ *Investigating Cellular Respiration*

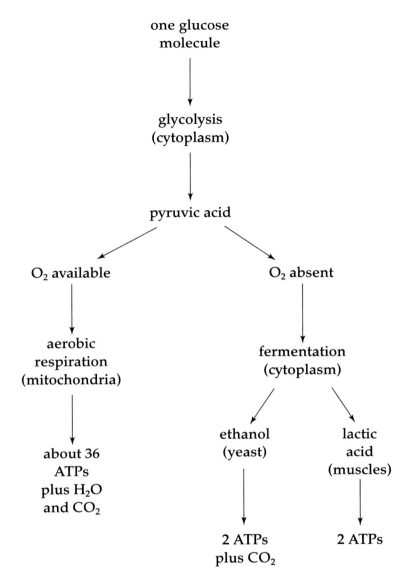

**FIGURE 4-3.** Steps of Aerobic and Anaerobic Respiration

Another method of anaerobic respiration is called **lactic acid fermentation**. During lactic acid fermentation, the pyruvic acid molecules from glycolysis are converted into **lactic acid**.

Some bacteria specialize in lactic acid fermentation. Lactic acid is produced when bacteria sour milk or ferment cabbage to form sauerkraut. The dairy industry uses lactic acid fermentation by bacteria and fungi to produce sour cream, cheese, and yogurt. Your muscle cells may use lactic acid fermentation when oxygen demands cannot keep up with supply (such as during heavy exercise).

The equation for lactic acid fermentation is

food molecules ⟶ pyruvic acid ⟶ ATP energy + *lactic acid* + heat

## Comprehension Check

The steps that occur during aerobic and anaerobic respiration are summarized in **Figure 4-3**. Refer to **Figure 4-3** and answer the following:

1. If **oxygen is available**, how many ATPs can a cell produce from **one** glucose molecule? _____

2. How many food molecules would a cell need to process to obtain the same amount of energy from **anaerobic** respiration? _____

3. Name the cell location where **glycolysis** (the first step of cell respiration) occurs. _____

4. Name the part of a cell in which **aerobic** respiration pathways occur. _____

5. Name the part of a cell in which **anaerobic** respiration pathways occur. _____

Check your answers with your instructor before you continue.

## ACTIVITY 6  ANAEROBIC RESPIRATION

**Note:**
Read through the instructions COMPLETELY before beginning.

1. Work in groups. Get the following supplies: **a pair of safety glasses, a tablespoon, two 400-ml beakers, a container of limewater, and a jar of sucrose.**

   In your classroom, you will find two **flasks** on a **warming plate**, labeled A and B. The flasks contain **water and yeast** cells.

   Each flask is closed with a stopper holding two pieces of tubing (see **Figure 4-4**). The longer piece of tubing is connected to an **air stone**.

2. Fill both beakers **half full** with **limewater**. Place the beakers on the table next to the warming plate. Insert the air stone from **Flask A into the first beaker** and from **Flask B into the second beaker**.

**Caution!**
Eye protection is required for the following procedures!

**Exercise 4** ■ *Investigating Cellular Respiration*

**FIGURE 4-4.** Setup for Fermentation Experiment

3. **Wearing eye protection,** gently remove the stopper from **Flask B and add one tablespoon of sucrose.**

   Replace the stopper. Record the time when you added the sucrose. _____

4. **Immediately** after adding the sucrose, remove the clamp from the air stone tube and **switch it to the vent tube. Switch the clamps on Flask 1 and Flask 2.**

5. Check on your experiment every **five minutes** for the next **15 minutes. Record** the results for **each five-minute period** in Table 4-4.

| TABLE 4-4 OBSERVATIONS OF YEAST FLASKS | | | | | | |
|---|---|---|---|---|---|---|
| OBSERVATIONS | EXPERIMENTAL FLASK A (WATER AND YEAST) | | | EXPERIMENTAL FLASK B (WATER, YEAST, AND SUCROSE) | | |
| | 5 MIN. | 10 MIN. | 15 MIN. | 5 MIN. | 10 MIN. | 15 MIN. |
| Color of liquid in yeast flask | | | | | | |
| Color of limewater in beaker | | | | | | |
| Presence of bubbles in yeast flask | | | | | | |

6. Did the results for Flasks A and B differ? Explain your answer by mentioning facts collected in this experiment.

7. What gas formed the bubbles observed in the flask(s)? _____

    **Explain your answer.**

| **Caution!** |
|---|
| Eye protection is required for the following procedures! |

8. **Wearing eye protection**, carefully remove the stopper from **Flask A** and sniff the released air. **Repeat** the process with **Flask B. Describe** the results.

# Exercise 4 ■ *Investigating Cellular Respiration*

## ✓ Comprehension Check

1. A yeast cell undergoing fermentation produced **six ATP molecules**. How many glucose molecules were used? _____ **Explain your answer.**

2. If the same cell was undergoing aerobic cellular respiration instead of fermentation, but using the same number of glucose molecules, how many ATP molecules would be produced? _____ **Explain your answer.**

Check your answers with your instructor before you continue.

### Note:

Lactic acid fermentation can only satisfy the body's short-term energy needs. If the oxygen shortage continues too long, lactic acid accumulates in your muscle tissue, lowering the pH. Eventually, this acidic environment inhibits the activity of enzymes that are needed for muscle function. Muscle fibers lose their ability to contract, a process called muscle fatigue.

The accumulated lactic acid can damage muscle proteins, producing the soreness and pain often associated with lengthy, strenuous exercise.

When oxygen is available again, the lactic acid is converted back to pyruvic acid and aerobic respiration can continue as before.

Self Test

1. A rock covered with green spots is placed on a screen in a test tube above some bromothymol blue solution; after several hours, the solution remains blue. What can you **conclude** from this experiment?

2. You and several other students observe that dead animals along the roadside often increase in size several days after they have been killed. In the spirit of scientific investigation, you collect the gas from inside one of these dead animals and bubble it through limewater. The limewater turns cloudy.

    a. What does the cloudy result indicate?

    b. If the animal is dead, how are these results possible? **Explain your answer in detail**.

3. Which produces the most energy in the form of ATP?

    a. Aerobic respiration
    b. Anaerobic respiration
    c. Alcohol fermentation
    d. All of the above produce the same amount of ATP energy but through different chemical reactions
    e. None of the above produces ATP

**Exercise 4** ■ *Investigating Cellular Respiration*

4. Carbon dioxide is passed into a solution of bromothymol blue indicator until the solution turns yellow. A sprig of elodea is then placed into this yellow solution. After a few hours in the sunlight, the yellow solution turns blue again. **Suggest an explanation** for the color changes observed in the bromothymol blue.

For Questions 5 through 8, use the following three answers. An answer may be used once, more than once, or not at all. For each question, **explain your answer**.

a. Process occurs only under aerobic conditions
b. Process occurs only under anaerobic conditions
c. Process occurs under both aerobic and anaerobic conditions

5. _____ Manufacturing of beer or wine

   Explanation:

6. _____ ATP production

   Explanation:

7. _____ Accumulation of lactic acid in a runner's muscles causes her to drop out of the race

   Explanation:

8. _____ Production of 36 ATPs, water, and carbon dioxide from one glucose molecule

   Explanation:

9. If an elephant's cells were to lose their ability to perform aerobic cellular respiration, what would most likely happen?

   a. The cells would switch over to anaerobic respiration and the elephant would be fine
   b. The cells would switch over to lactic acid fermentation and the elephant would survive but would have achy muscles
   c. The cells would switch over to alcohol fermentation and the elephant would survive in a drunken stupor
   d. The cells would die because the elephant's energy needs could not be met using anaerobic metabolic pathways
   e. The cells could live on stored energy for an extended period, and cellular respiration would not be necessary for several years

   **Explain your answer.**

EXERCISE 5

# The Skin: An Example of an Organ System

## Objectives
After completing this exercise, you should be able to:
- list and explain several features present in the epidermis that provide for protection of the body
- explain how each skin layer is different from the others
- identify each structure in the dermis and explain its function
- list and explain at least four examples of how deposits of fats and oils are useful to plants, animals, and microorganisms
- discuss several adaptations of epithelial tissues that demonstrate the relationship between structure and function
- discuss the relationship between the density of touch receptors in various body locations and touch sensitivity
- apply your knowledge of the special features of the skin to practical situations.

## CONTENT FOCUS

Within our bodies, we have approximately 100 trillion cells. As you observed in previous exercises, not all cells have the same **structure.** In the bodies of multicellular animals such as humans, not all cells have the same **function.**

A **tissue** is a **group of similar cells that perform a specific function.** Cells are organized into **four major tissue types:**

| | |
|---|---|
| **Epithelial Tissue** | Lines all inner and outer body surfaces; covers organs and body cavities |
| **Muscle Tissue** | Contracts to produce movements |
| **Connective Tissue** | Joins and supports other tissues |
| **Nervous Tissue** | Senses stimuli; transmits signals around the body |

**Tissues group together to form the organs** of the body. All of these tissue types are present in the **skin,** the body's largest organ.

To examine tissue function, you will take a closer look at the **design of your arm,** beginning with the outer epidermis and working your way through the underlying tissues to the supporting skeleton.

# ACTIVITY 1  SKIN: THE OUTER PROTECTIVE LAYER

1. Work in **groups**. Set up a **dissecting microscope.**

2. Place your hand, **palm-side down,** on the stage of the microscope. **Examine your skin** closely, beginning with the **lowest** magnification and gradually zooming up to the **highest.**

   **Record** your skin **observations** (in detail).

3. **Remove your hand from the microscope stage.** Get a **dropper bottle of water.**

   While holding your hand level in front of you, **palm-side down (NOT under the microscope),** place a drop of water on the back of your hand.

   **What happens** to the water droplet?

   Gently **tilt** your hand. **What happens** to the water droplet now?

   Can liquids easily penetrate the skin? _____

   Is this one of the **protective functions** of the skin? _____ **Explain your answer**.

4. The **waterproofing** quality of the skin is due to the presence of the **protein keratin** in epithelial cells.

   The **keratinized** cells of the epidermis also protect the underlying tissues from **mechanical injuries and abrasions.** On parts of the body where **abrasion is most common,** the epidermal layer tends to be **thicker.**

**Exercise 5** ■ *The Skin: An Example of an Organ System*

5. The skin also provides protection from exposure to **ultraviolet (UV) radiation.** Pigment producing cells in the epidermis manufacture the **protein melanin** that blocks penetration of UV rays and protects underlying cells from damage.

   Most people have about the same number of melanin-producing cells, called **melanocytes. Dark-skinned** individuals, however, produce **more and darker melanin** than fair-skinned individuals. Melanin production is also **stimulated by exposure to sunlight.**

6. Problems with melanin production can result in different disorders.

   **Albinos** have a normal number of melanin-producing cells, but the cells do not synthesize melanin. For this reason, albinos have no pigment in their skin, eyes, and hair. In the condition **vitiligo,** melanocytes die, causing patches of skin to lose their coloration. These light spots in the skin are often surrounded by skin with normal pigmentation.

## Comprehension Check

1. Which of the following areas of the body probably have a thick epidermal layer? **(Circle ALL correct answers.)**

   a. inside of the cheeks
   b. palms of the hands
   c. soles of the feet
   d. scalp
   e. elbows
   f. abdomen

2. The thick layers of keratinized cells mentioned above form highly distinctive patterns of **ridges and whorls** in some body locations.

   We call these patterns _____.

3. To raise some money for your tuition, you take a summer job working construction. After several weeks, you notice **calluses** have developed on your hands. What is a **callus? Why did the calluses form?**

4. Based on the information in **Activity 1**, what causes a **suntan?**

5. **Skin cancer** is increasing in our population. **(Circle one answer.)** You have better protection from exposure to the damaging effects of UV rays if you have **dark/fair** skin.

 Check your answers with your instructor before you continue.

## ACTIVITY 2  THE EPIDERMIS

Preparation

By looking at **Figure 5-1**, you can see that the skin is made up of several layers: the **epidermis** (the outermost layer), the **dermis** (the middle layer), and the **hypodermis** (the innermost layer).

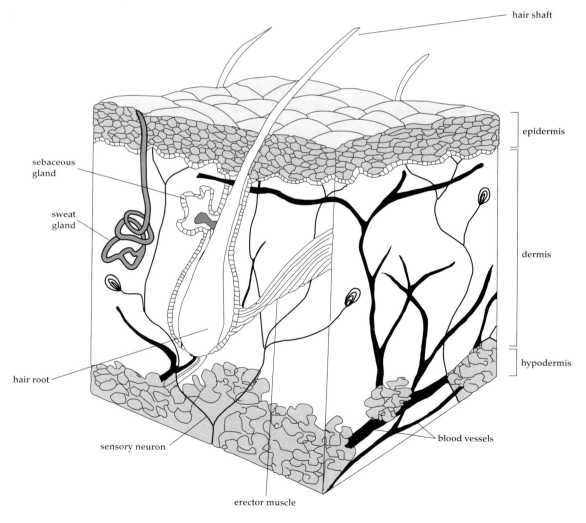

**FIGURE 5-1.** Layers of the Skin

**Exercise 5** ■ *The Skin: An Example of an Organ System*

1. Look at **Figure 5-2** and note that the **epidermis** consists of many layers of **epithelial cells.**

   You can also see the thickness of the epidermis in comparison to the other skin layers by looking at the top of the diagram in **Figure 5-1**.

   The protective layer of keratinized cells at the **surface** of the epidermis is **repaired by rapidly dividing cells** at the **base** of the epidermis. Newly formed cells are gradually pushed upward to replace lost and damaged cells in the outer layer. In some individuals, excessive shedding of cells occurs in the scalp area.

   These discarded scalp cells are referred to as _____.

## Comprehension Check

1. On **Figure 5-2**, draw an **arrow** showing the **direction of cell replacement** in the epidermis.

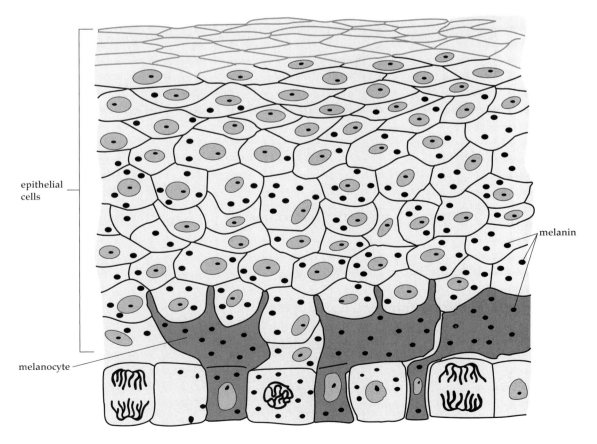

**FIGURE 5-2.** The Epidermis

2. Dividing cells have **distinctly visible chromosomes** within them (these appear as dark lines).

   Using a **colored pencil or highlighter**, color in **two** cells that are dividing.

3. **(Circle one answer.)** Actively dividing cells can be found in the **top/middle/bottom** layer of the epidermis.

4. On **Figure 5-2**:

   Place an **"O"** inside a cell that would be one of the oldest cells in the epidermis.

   Place a **"Y"** inside a cell that is one of the youngest cells in the epidermis.

   Place a **"B"** inside a cell close to a blood vessel.

5. If the prefix **"epi"** means *above*, where do you predict the **dermis** of the skin will be located?

Check your answers with your instructor before you continue.

## ACTIVITY 3 — THE DERMIS

1. By looking at **Figure 5-1, circle the correct answer** in each of the following statements.

   The dermis is located **above/below** the epidermis.

   The dermis is **thicker/thinner** than the epidermis.

   Hair roots are located in the **dermal/epidermal** layer of the skin.

2. Take a look at the epidermal and dermal layers in **Figure 5-1**.

   Are any **blood vessels** present in the **epidermis?** _____

   If you cut **just** your epidermis, will the cut bleed? _____

**Exercise 5** ■ *The Skin: An Example of an Organ System*

3. Blood vessels carry nutrients and oxygen to body cells.

   **(Circle one answer.)**

   Epidermal cells at the **base** of the epidermis have **excellent/fair/poor** access to oxygen and nutrients.

   Epidermal cells at the **surface** of the epidermis have **excellent/fair/poor** access to oxygen and nutrients.

   The farther epithelial cells are removed from good access to nutrients and oxygen, the **better/worse** their chances of survival.

4. **Blood vessels** near the root area supply the hair with **oxygen and nutrients** needed for growth.

   **Melanin granules** in the center of the hair shaft give the hair its color.

   Notice that **sebaceous glands** are located along the hair shaft. Oily secretions from the sebaceous glands **lubricate** the hair and skin, help keep the skin from **drying out**, and **inhibit growth of bacteria**.

✓ Comprehension Check

1. The sale of **hand lotion** is a multimillion dollar market in the United States. Using your knowledge of how the epidermis is lubricated, explain why hand lotion is such a big-selling item.




2. Explain why the skin of a person's scalp is often more oily than the skin of his or her arm.

**Check your answers with your instructor before you continue.**

# ACTIVITY 4      HAIR TODAY, GONE TOMORROW

*Preparation*

Millions of hairs are scattered over the body. The average scalp has about 100,000 hairs and a man's beard has another 30,000.

1. Refer back to **Figure 5-1**. You will see that a hair consists of two parts, the **shaft** and the **root**. The shaft portion of the hair begins in the **dermis** and extends **outside** the surface of the skin.

   The **root** is found **only within** the dermis and is surrounded by several layers of cells. Collectively, this structure is called a **hair follicle**.

2. Sit with your **eyes closed** while your partner runs his or her hand over the top of your head, **gently** touching the **hairs**.

   Can you tell when you are being touched? _____

   What **section** of the hair was just touched? _____

   A knot of **sensory nerve endings** is wrapped around the **root of each hair**.

   **Moving a hair activates these nerve endings,** so that your hairs function as sensitive touch receptors.

3. The outer part of the shaft contains heavily **keratinized** cells, which appear roughened and scalelike under a microscope.

   Hair pigment is made by **melanocytes** at the base of the hair follicle. Several types of pigments combine to produce the various hair colors.

   **(Circle one answer.) Brown/red/black/blond** hair has the most melanin present.

4. Living cells **divide** at the base of the follicle, causing your hair to grow. **Blood vessels** provide follicle cells with the oxygen and nutrition needed for growth.

5. Refer back to **Figure 8-1**. You can see that each hair is connected to a **tiny erector muscle** that **automatically** raises the hair when you are cold.

   In animals with fur coats, raised body hairs **hold a layer of warm air** next to the skin, insulating them from the cold. Humans have too few body hairs for this response to help keep us warm, but we are like other animals in our response to changing temperatures.

   Contraction of the muscles that raise body hairs causes _____ to appear on the surface of the skin.

**Exercise 5** ■ *The Skin: An Example of an Organ System*

## Comprehension Check

1. Which of the following skin features can be found in the **dermis**? **(Circle ALL correct answers.)**

   a. melanin
   b. blood vessels
   c. nerve endings
   d. sebaceous glands
   e. erector muscles
   f. hair follicle
   g. hair shaft
   h. highly keratinized cells
   i. sweat glands

2. If you get a paper cut that penetrates to the **dermis,** will the cut bleed? _____ **Explain your answer.**

3. If the paper cut penetrates **only** the **epidermis,** will the cut **hurt?** _____ **Explain your answer.**

4. Burns and other injuries sometimes causes the epidermis to separate from the dermis.

   When the space fills up with fluid, we call this a _____.

**Check your answers with your instructor before you continue.**

## ACTIVITY 5          BELOW THE SKIN (THE HYPODERMIS)

1. Beneath the skin is a layer of **connective tissue** with many fat cells. (See **Figure 8-1.**) This layer is called the **hypodermis** ("hypo" meaning **under** the dermis) or **subcutaneous** layer ("sub" meaning *under* and "cutaneous" referring to the skin).

   Perform the following experiment using a **"blubber mitten"** to discover one of the functions of the subcutaneous fat layer.

2. Work **individually.** Get the following supplies: **one blubber mitten and one empty mitten.**

   The two mittens are designed to simulate how the subcutaneous layer of the skin would function **with and without fat cells.**

   In this experiment, you will place one mitten on each hand and immerse both hands in an **ice-water bath.**

3. **Before proceeding,** form a hypothesis about the **effect of water temperature** on each of your hands. **Record** your hypotheses below.

| HYPOTHESES FOR BLUBBER MITTEN EXPERIMENT |
| --- |
|  |

4. Go to the ice-water bath and **perform your experiment. Record your results** below.

| RESULTS OF BLUBBER MITTEN EXPERIMENT ||
| --- | --- |
| WITHOUT FAT | WITH FAT |
|  |  |

5. From the results of your experiment, what **conclusions** can you draw about the effect of fatty tissue in the subcutaneous layer of the skin?

**Exercise 5** ■ *The Skin: An Example of an Organ System*

## Comprehension Check

1. Which of the two mittens was the **control** in your experiment? _____

2. Why was it **necessary** to use a **control mitten** (instead of putting your **bare hand** in the ice-bath)?

3. **Based on the results** of your experiment, give **one function** for subcutaneous fat tissue.

4. Again, **referring to the results** of your experiment, can you suggest **two reasons** why some animals eat a lot in the fall season?

    a.

    b.

5. You are probably familiar with the fact that, when you **mix oil and water,** the oil will float to the surface. Do you think that significant **fat** deposits, such as those found in marine mammals, contribute to their buoyancy? **Explain your answer**.

6. On an ocean field trip with your marine biology class, you make a collection of **phytoplankton (tiny marine plants).** When you look at these plants through the microscope, you notice that many of them have **oil droplets** inside their cells.

    Suggest **two ways** in which oil droplets might be helpful to the phytoplankton.

    a.

    b.

7. Often **injections** of medicine are administered into the **subcutaneous layer** of the skin.

    The familiar term _____ **needle** refers to the location in the skin where the needle is inserted.

**Check your answers with your instructor before you continue.**

# ACTIVITY 6      THE SENSORY FUNCTION OF THE SKIN

Preparation

The skin acts as an interface between the body and the outside environment. Information about several aspects of the outside world is acquired through activation of sensory receptors in the dermis and epidermis of the skin.

Sensory neurons process and transmit incoming information to the central nervous system (the brain and spinal cord). Neurons vary in function and appearance, but all share a similar three-part structure: the cell body, the axon, and the dendrites (see **Figure 5-3**).

**FIGURE 5-3.** Structure of a Typical Neuron

**Exercise 5** ■ *The Skin: An Example of an Organ System*

Every neuron has a **cell body,** the region of the cell that contains the nucleus and a variety of other organelles necessary for cell metabolism. Outside stimuli are received by **dendrites** in the skin and other sense organs. Incoming signals from these dendrites are transmitted toward the cell body and continue into the **axon.** Axons relay outgoing messages from one neuron to other neurons or to various tissues and organs in the body.

Dendrites in the skin are specialized to receive different types of sensory input, such as **touch, temperature, pressure, pain, vibration, and proprioception** (the sense that tells you the current position of your body and limbs). In the following experiment, you will be making observations about the distribution of touch receptors in the skin.

1. Work **with a partner.** Get the following supplies: **one set of touch calipers.**

   The set of touch calipers consists of a series of corks. Each cork has two pins inserted so that the blunt ends are protruding. The pins are spaced at varying distances from each other (**1mm, 2mm, 4mm, 8mm, 16mm, and 32 mm apart**).

2. Form a hypothesis about the relative sensitivity of various body locations. For the six body locations mentioned in **Table 5-1**, list them in order from the area that you think will be **least sensitive** to the area you think will be **most sensitive.**

| TOUCH SENSITIVITY HYPOTHESIS |
|---|
|  |

3. Perform the following set of tests on your partner, then reverse roles and have your partner use the same tests to check your touch response sensitivity.

   Your partner should be seated, with his/her **eyes closed.**

   **For each touch given, the subject will report whether he/she feels one or two pins touching the skin.**

4. Begin with the **32 mm caliper.** Lightly touch the skin on the back of the test subject's neck.

   Repeat the process with the remaining calipers in the set **(16 mm, 8 mm, 4 mm, 2 mm, and 1 mm) in any order you wish.**

   **For each test, record whether the person is able to feel two pins touching the skin.**

If the test subject felt **two pins,** record **yes.** If the test subject felt **one pin only,** record **no.**

Record these results in **Table 5-1**.

5. Repeat the same procedures at each of the following locations:
    - back of the hand
    - palm of the hand
    - inside of the forearm
    - tip of the nose
    - cheek
    - tip of the forefinger

6. **Reverse roles** and have your partner repeat the same tests to check your touch response sensitivity. Record the results in **Table 5-1**.

### TABLE 5-1
### TOUCH SENSITIVITY IN VARIOUS BODY LOCATIONS

**TEST SUBJECT 1**

| Body Location | 32 mm | 16 mm | 8 mm | 4 mm | 2 mm | 1 mm |
|---|---|---|---|---|---|---|
| Back of neck | | | | | | |
| Back of hand | | | | | | |
| Palm of hand | | | | | | |
| Inside of forearm | | | | | | |
| Tip of nose | | | | | | |
| Cheek | | | | | | |
| Tip of forefinger | | | | | | |

**TEST SUBJECT 2**

| Body Location | 32 mm | 16 mm | 8 mm | 4 mm | 2 mm | 1 mm |
|---|---|---|---|---|---|---|
| Back of neck | | | | | | |
| Back of hand | | | | | | |
| Palm of hand | | | | | | |
| Inside of forearm | | | | | | |
| Tip of nose | | | | | | |
| Cheek | | | | | | |
| Tip of forefinger | | | | | | |

**Exercise 5** ■ *The Skin: An Example of an Organ System*

The **greater the number of touch receptors** in a body location, the **more sensitive** that location will be to touch stimuli. The reason for the increased sensitivity lies in the distribution of the receptors. If the pin points touch **two receptors that are adjacent** to each other, the pins will **activate both receptors together** and it will feel as though you have been touched by **only one pin.**

If the pins touch two receptors that are **not adjacent,** however, each will **fire separately** and you will feel the touch of **both pins**. The ability to feel the two pins is called **"two-point discrimination."**

In locations with a high density of touch receptors, therefore, the skin is much more sensitive to touch and much better at providing information about the number of touches (one pin or two pins). This type of touch sensitivity allows a surgeon or mechanic to perform delicate procedures working by touch alone.

The smallest pin distance when the test subject had effective two-point discrimination indicates the body location with the highest density of touch receptors.

The higher the number of touch receptors in a specific location, the greater the probability the pins can stimulate two non-adjacent touch receptors.

### ✓ Comprehension Check

1. Based on the results in **Table 5-1**, rank all areas tested in terms of their relative sensitivities to touch. List the body locations from the area of **least sensitivity** to the area of **highest sensitivity.**

    Subject 1:

    Subject 2:

2. Which of the tested body locations has the **greatest density** of touch receptors? **Explain your answer**.

3. Were there differences in touch sensitivity between you and your partner? If so, what differences did you observe?




4. Would you expect a decrease in touch sensitivity for each of the following conditions? **Explain each answer**.

   a. calluses on the palms of your hands:



   b. scar tissue:



   c. thick subcutaneous fat layer:



5. People who are visually impaired are able to read books and other printed materials by using the Braille system of writing. In the Braille system, words are represented by patterns of raised dots. Braille readers touch the dot patterns with their fingertips to read the text.

   How does the density of touch receptors in the fingertips make the Braille system possible?

**Check your answers with your instructor before you continue.**

**Exercise 5** ■ *The Skin: An Example of an Organ System*

Self Test

Fill in the blank with the choice that is **most appropriate** to describe the function of each skin structure. Answers can be used **only once**.

a. keratin
b. melanin
c. erector muscle
d. epithelial cell
e. epidermis
f. hair follicle
g. blood vessels
h. subcutaneous layer
i. sebaceous gland
j. dermis

1. _____ Produces oil to lubricate the hair and skin.

2. _____ The layer of skin containing many blood vessels and nerves.

3. _____ Type of protein deposits in the skin that form fingerprints.

4. _____ Type of dead cell that flakes off as dandruff.

5. _____ When the hairs on the back of a dog stand up, this tissue is responsible.

6. _____ Location of fatty tissue that insulates and protects the body.

7. _____ Accumulation of this protein helps protect your skin from the sun's rays.

8. _____ The outermost layer of the skin.

9. We often hear claims made by cosmetic companies that their **lotions** will **"make your skin young again."** Based on the information you have gained in this exercise, do you think that this is an accurate statement? **Explain your answer.**

10. You are determined to quit smoking, so you decide to try a brand of nicotine "patch." Where is the **closest location** for the nicotine to enter your circulatory system? **Explain your answer**.

11. In reference to the nicotine patch: Even though it would not be as visible, why would it be a bad idea to place the patch on the sole of your foot? (Answer in terms of the ability of the patch to **function.**)

12. *Challenge Question!* Burns are classified according to how many skin layers are destroyed. A **first-degree burn** only affects the outer layers of the epidermis. A **second-degree burn** destroys the epidermis and penetrates the dermis. **Third-degree burns** destroy skin tissues down to and including the subcutaneous layer.

    A nurse who works in the burn ward of a local hospital notices that patients who have second-degree burns suffer more pain than patients who have third-degree burns.

    Using your knowledge of the skin, **explain** why the more severe burn causes less pain.

# EXERCISE 6

# The Musculoskeletal System

## Objectives
After completing this exercise, you should be able to:
- identify three different types of muscle tissue using the microscope
- discuss the functions and body location of each type of muscle tissue
- explain the effects of muscle fatigue on arm muscle activity
- identify the major structures of a long bone and explain the function of each
- discuss the contribution of minerals and protein fibers to bone strength and flexibility
- apply your knowledge of bone and joint structure to practical examples of skeletal support and movement.

## CONTENT FOCUS

In previous exercises you looked at the protective outer covering of the body. Now you will move beneath the skin to examine muscles and bones.

Consider an Olympic sprinter at the beginning of a race. The moment the starting gun fires, muscles contract, the skeleton responds, joints allow for changes in body position, and tissues all over the body work together to coordinate forward movement. Tasks as simple as sitting, standing, walking, or taking notes in class all require similar coordination of muscles, bones, and joints.

## ACTIVITY 1                               BETWEEN THE SKIN AND BONES

1. If you were to take a **small piece of steak** and make a wet mount of the **muscle tissue,** you would see alternating light and dark bands. These bands or **striations** are composed of two proteins **(actin and myosin)** which are involved in **muscle contraction.**

2. To observe bands of actin and myosin, get a **prepared slide of skeletal muscle** (also called **"striated,"** meaning striped, muscle).

3. Observe the slide with the **compound microscope** on **high power.** Refer to the **photograph** in **Figure 6-1** to locate the following structures on your slide:

   a. **one** muscle cell (also called a muscle **fiber**)
   
   b. striations
   
   c. nuclei of **one** muscle cell

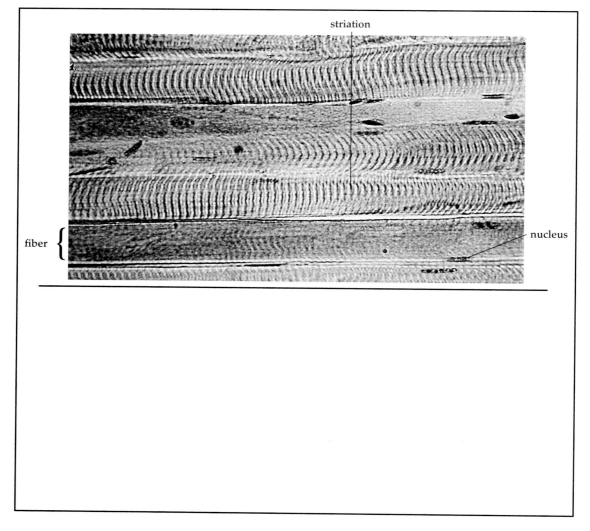

**FIGURE 6-1.**   Skeletal Muscle Tissue

# Exercise 6 ■ The Musculoskeletal System

4. In the empty box in **Figure 6-1**, make a drawing of several skeletal muscle cells **as they appear in your microscope.**

5. **Label** the **striations, nuclei, cytoplasm, and cell membrane** of **one** skeletal muscle cell on your **drawing.**

 Check your answers with your instructor before you continue.

6. The **function** of skeletal muscle is to bring about **voluntary** movement. Skeletal muscles are under your conscious control. **Wiggle the fingers** of your left hand. You are using skeletal muscles!

7. There are **two other types of muscle tissue** that have **different functions** in the body. **Smooth muscle** allows for **involuntary** movements—that is, you have little control over its function.

   Observe the **two drawings** of smooth muscle in **Figure 6-2**.

   The protein fibers of smooth muscle cells are **not** arranged in bands, so **striations are not visible** in this tissue. Smooth muscles are found in the **digestive tract, blood vessels,** various **glands,** and many other locations where **automatic responses** are required.

8. **Add labels** for the following structures to the drawing of **individual smooth muscle cells** in **Figure 6-2**: **cell membrane, nucleus, and cytoplasm.**

9. **Cardiac muscle** is found **only** in the **heart**. It is responsible for your heartbeat. Refer to the photograph and drawing in **Figure 6-3**.

   Are cardiac muscle cells **striated** (similar to skeletal muscle cells)? _____

10. Contraction of all cardiac muscle fibers **must be coordinated** in order to maintain a regular heartbeat.

    To accomplish this coordination, cardiac muscle cells are tightly connected into a communication network that sends the signal to contract from cell to cell. These cell junctions are called **intercalated discs**.

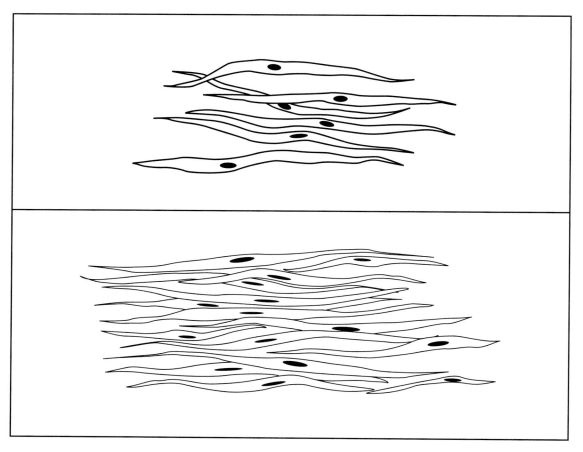

**FIGURE 6-2.** Smooth Muscle Tissue

## ✓ Comprehension Check

Now that you are familiar with the **three different types of muscle tissue,** answer the following:

1. The erector muscles that **raise hairs on your arm** consist of _____ muscle tissue.

2. When you **wink** your eye, what type(s) of muscle tissue(s) is(are) in use?

# Exercise 6 ■ The Musculoskeletal System

**FIGURE 6-3.** Cardiac Muscle Tissue

3. When you are overheated, extra blood is directed to the skin for cooling. What type(s) of muscle tissue is(are) involved?

4. During **labor and childbirth,** what type(s) of muscle tissue(s) is(are) contracting? **Explain your answer**.

5. List **two** locations of smooth muscle in the body that were **not** mentioned in this exercise.

   _____    _____

6. During a heart attack, _____ **muscle** may be damaged.

Check your answers with your instructor before you continue.

## ACTIVITY 2 — MUSCLE FATIGUE

*Preparation*

Muscles that are actively contracting become weaker as time passes. This process is known as **muscle fatigue.** Muscle fatigue can be explained by several factors that occur simultaneously as your body works. These include:

- lack of ATP to meet energy needs
- insufficient oxygen for cell respiration
- depletion of energy reserves in the muscle cells
- accumulation of metabolic wastes (such as lactic acid)

1. Work in groups. During this experiment, you will be supporting a **heavy** book on your open palm with your arm completely extended. Which arm can support the book longer? **Form a hypothesis** about the ability of the muscles in your right and left arm to support the book and **record** your hypothesis below.

| HYPOTHESIS FOR MUSCLE FATIGUE EXPERIMENT |
|---|
|   |

Check your hypothesis with your instructor before you continue.

**Exercise 6** ■ *The Musculoskeletal System* 103

2. For your muscle fatigue experiment, one member of the group will act as a **timekeeper**, the second will **record** the experimental results, and the third will be the **test subject**.

> **Note:**
> The person holding the book should not know how much time has expired until the entire experiment is completed.

3. Get an **extremely heavy book.**

   a. Place the book on your open hand.

   b. With the book on your hand, extend your arm fully with the palm up. **Do not bend your elbow.**

   Your arm should not be braced against your body.

   c. Record the length of time you can hold your arm extended. **Record** the results in **seconds** under **Trial 1** in **Table 6-1**.

   d. Rest your arm for **5 seconds** and repeat the experiment for **Trial 2.**

   e. Rest your arm for **5 seconds** and repeat the experiment for **Trial 3.**

   f. Rest your arm for **5 seconds** and repeat the experiment for **Trial 4.**

### TABLE 6-1
### RESULTS OF MUSCLE FATIGUE EXPERIMENT

| Trial Number | Time Arm Held Extended (Sec) Right Arm | Time Arm Held Extended (Sec) Left Arm |
|---|---|---|
| 1 | | |
| 2 | | |
| 3 | | |
| 4 | | |

4. **Graph** your experimental results in **Figure 6-4**. Plot **time** on the **Y axis**.

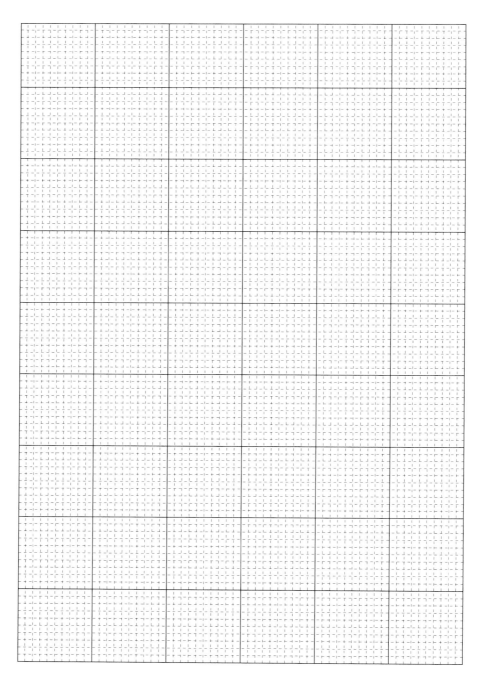

**FIGURE 6-4.** Comparison of Muscle Fatigue in Right and Left Arm

**Exercise 6** ■ *The Musculoskeletal System* **105**

5. Did the experimental results **support** your hypothesis? _____

   **Explain your answer**, mentioning **facts** collected during your experiment.

6. Did you see evidence of **muscle fatigue? Explain** your answer.

7. To have full confidence that your conclusions about muscle fatigue are accurate, **what changes would you recommend** for the experimental design?

**Check your answers with your instructor before you continue.**

## ACTIVITY 3　　　　　　　　　　　　　　　　　　　　SKELETAL SUPPORT

Preparation

Bone is a type of **connective tissue.** Although bone may not superficially resemble other body tissues, it **is** composed of **living cells.**

Bone cells produce a rigid material that forms the structure generally associated with bones. This mineral **"matrix"** in humans is composed of **calcium phosphate.**

Interwoven throughout the rigid mineral matrix are **protein fibers** that add **strength and flexibility.**

**Compact bone** is strong and densely packed. In compact bone, the cells are organized into vertical structures that resemble **drinking straws.** If you try to crush a straw by pushing on both ends, it will be hard to do. In a similar manner, bones resist the force of gravity and support your body weight.

In contrast to compact bone, **spongy bone** consists of a loose arrangement of thin plates of bone similar to the appearance of a sponge. **Spongy bone is lighter** than compact bone, **reducing the weight of the skeleton** and making it easier for the muscles to move the bones.

Bone strength is a result of compact and spongy bone working together. **Compact bone** can only resist forces applied **along the length** of the bone, while **spongy bone** helps resist stresses **applied from many directions** (such as a kick to the side of your leg).

The inside of a bone contains **bone marrow.** The central cavity of the bone **shaft** contains **yellow marrow,** which functions as a fat-storage area. At the **ends of the long bones,** gaps in the spongy bone contain **red marrow,** which produces all types of **blood cells.**

Bone cells, as with cells of all tissues, require delivery of oxygen and nutrients and removal of waste materials. Although you cannot see them without magnification, **channels throughout the bone** provide passageways for a network of blood vessels and nerves to serve these cells.

1. Work in groups. Get a **long bone** that has been cut open lengthwise.

   Set up a **dissecting microscope** for your observations.

2. **Examine** the bone with and without the microscope.

   On the basis of your **observations,** label the following structures in **Figure 6-5**: **compact bone, spongy bone, shaft, marrow cavity, the location of red marrow, and the location of yellow marrow.**

3. **(Circle one answer.)** Most blood cells are produced in the **shaft/ends** of the bone.

   **Explain** your answer.

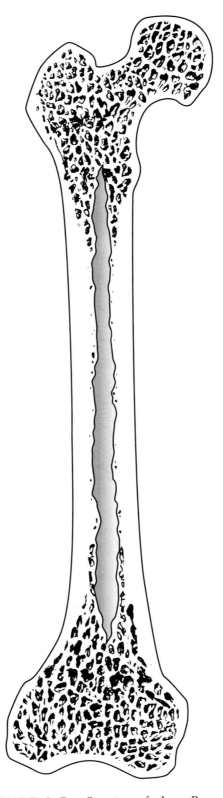

**FIGURE 6-5.** Structure of a Long Bone

4. (Circle one answer.) In situations of long-term starvation, **red marrow/yellow marrow** would probably decrease first.

    **Explain** your answer.

5. Examine a **human pelvis,** which is composed of several attached bones.

    Based on the function of the pelvis and its position in the skeleton, form a hypothesis about the relative **amounts of compact and spongy bone** present in the bones of the pelvis. **Record** your hypothesis below.

    | HYPOTHESIS: COMPACT VS. SPONGY BONE IN THE PELVIS |
    |---|
    |  |

6. **Explain** the line of reasoning you followed to develop your hypothesis.

7. Suggest a method you could use to **test** your hypothesis.

**Check your answers with your instructor before you continue.**

**Exercise 6** ■ *The Musculoskeletal System*

8. *Challenge Question!* In many societies, heavy bundles are balanced on people's heads, rather than carried in their arms or on their backs (see **Figure 6-6**).

    **Explain** how **compact and spongy bone** each contribute to the ability of the skeleton to support the load.

**FIGURE 6-6.** Supporting a Load on Your Head

# ACTIVITY 4

# WEAK AND STRONG BONES: CLASSROOM DEMONSTRATION

Bone is about 75% **minerals (calcium phosphate)** and 25% **protein fibers (mostly collagen)**. These materials, **acting together,** give bone its strength and flexibility.

1. Your instructor has prepared several **chicken bones** that have been subjected to **three different treatments.** The bones will be used to demonstrate how **minerals and protein fibers** in bone tissue contribute to strength and flexibility.

    **Dish A**   Contains cleaned chicken bones that were soaked in acid for several days. Under these conditions, **minerals** are removed, but the **protein** content does not change.

    **Dish B**   Contains cleaned chicken bones that were baked in a very hot oven. Under these conditions, the **protein fibers** are destroyed, but the **mineral** content does not change.

    **Dish C**   Contains chicken bones that were cleaned and dried.

2. Which dish contains the **control** bones? _____

3. For each statement, **circle one answer**.

    The acid-treated bones are **more/less flexible** than the control bones.

    The acid-treated bones are **harder/softer** than the control bones.

    **Explain your answer** to both statements.




4. For each statement, **circle one answer**.

    The baked bones are **easier/harder** to **break in half** than the control bones.

    The baked bones are **easier/harder** to crush **into powder** than the control bones.

    **Explain your answer** to both statements.

**Exercise 6** ■ *The Musculoskeletal System*  111

### ✓ Comprehension Check

1. In the condition known as **rickets,** the leg bones bend under the body's weight. Using your knowledge of the role of **minerals and protein fibers** in bone structure, explain one possible cause of rickets.

2. In the condition **osteoporosis,** the spongy bone of the pelvis and femur (long bone of the thigh) can be affected by severe loss of bone mass. Using your knowledge of bone structure, explain why a person with this disease would be more susceptible to hip fractures.

**Check your answers with your instructor before you continue.**

## ACTIVITY 5 — A JOINT VENTURE: MOVING THE BONES OF THE SKELETON

*Preparation*

Since bones are rigid, the skeleton can only change position through the action of joints. Most joints of the body are **synovial joints.** At synovial joints, the adjacent bones are separated by a fluid-filled capsule. Some joints, such as the knee, give us more problems than others. The problems originate because of weakness in the joint supports.

In the hip and shoulder joints, a **ball** on the end of one bone fits securely into a **socket** on another bone. The knee joint, in contrast, is held together only by tough bands called **ligaments.** The ligaments stretch like thick rubber bands between the ends of the thigh and shin bones, pulling them toward one another.

The **collateral ligaments** support the joint on the right and the left sides. The **cruciate ligaments** cross inside the knee from front to back, preventing the knee from overextending.

1. Work in groups. Get a **knee model.**

2. Locate the **two collateral and two cruciate ligaments.** Ligaments form attachments **between bones. Label** the **four support ligaments** on **Figure 6-7**.

3. The knee joint connects the **femur** (thigh bone) and the main support bone of the lower leg, the **tibia** (shin bone). The smaller lower leg bone, the **fibula,** is not part of the knee joint. Although the fibula does not help support body weight, it is an important surface for muscle attachment and provides stability to the ankle joint.

   **Label** the three leg bones on **Figure 6-7**.

**FIGURE 6-7.** Synovial Joint of the Knee

4. The **patella** (kneecap) protects the joint and prevents the knee from bending in the wrong direction.

5. The **meniscus,** a thick pad between the femur and tibia, **improves the fit** between the bone ends, making the joint more stable. Locate the **meniscus** on the knee model and label it on **Figure 6-7**.

6. Muscles and bones are connected by **tendons.** The ligaments and tendons of the knee form a **capsule** around the joint (see **Figure 6-8**).

7. No matter how smooth the ends of the bones may look, their surfaces are still uneven. When placed close together, these uneven surfaces rub against one another, producing heat, friction, and wear. In joints, the ends of the bones are covered with a **cushioning layer** of **articular cartilage.**

8. A membrane that lines the joint capsule, called the **synovial membrane,** produces **fluid** that lubricates the joint (called **synovial fluid**), reducing friction.

9. Although they are not present on the model, the knee joint also has many fluid-filled sacs called bursae (singular **bursa**).

    These tiny bags of lubricating fluid reduce rubbing in areas where bones come in contact with ligaments, muscles, and skin.

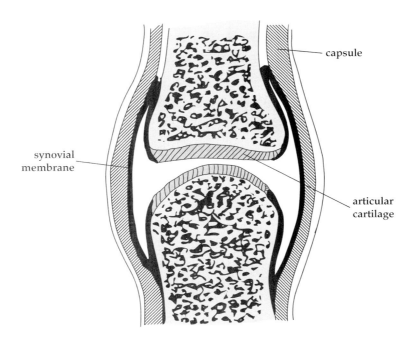

**FIGURE 6-8.** Interior of a Synovial Joint

## Comprehension Check

Fill in the blank with the **most appropriate** answer. Answers can be used **only once**.

a. meniscus
b. collateral ligament
c. articular cartilage
d. synovial membrane
e. synovial fluid
f. cruciate ligament
g. bursae
h. patella

1. _____ Covers the rough ends of bones, creating a smooth surface.

2. _____ A frequent football injury occurs when a player is hit on the side of the knee, tearing this structure.

3. _____ When this structure tears, there is nothing to prevent the knee from extending too far forward.

4. _____ If this structure, which provides a cup for the ends of the long bones, is damaged, the bones may slip out of position in the joint.

5. _____ Excess "fluid on the knee" is produced by this structure.

6. _____ Repetitive movement may lead to inflammation of these bags of cushioning fluid, producing swelling and pain.

**Check your answers with your instructor before you continue.**

**Exercise 6** ■ *The Musculoskeletal System*

Self Test

Fill in the blank with the choice that is **most appropriate** to describe the function of each in the body. Answers can be used **MORE THAN** once.

a. spongy bone
b. skeletal muscle
c. cardiac muscle
d. yellow marrow

e. red marrow
f. smooth muscle
g. compact bone
h. shaft

1. _____ A storage tissue for fat in bones.

2. _____ Tissue that produces blood cells.

3. _____ Densely packed tissue found at the outside of an arm bone.

4. _____ When the hairs on the back of a dog stand up, this tissue is responsible.

5. _____ When an emergency room doctor use an electrical shock to start someone's heart, they are stimulating this type of muscle.

6. _____ When you touch a hot stove and jerk your hand away, this muscle type is contracting in your arm.

7. _____ This type of muscle moves the leg at the knee joint.

8. _____ Skeletons of birds adapted for flight would have more of this type of bone.

9. _____ This part of a long bone will be longer in taller people.

10. _____ If you placed an arm bone in a vise and applied pressure to both ends, this type of bone would resist breaking.

11. You and a friend are at the gym working out. Your friend is upset because she can't do as many repetitions of a bench press in her third set as she did in her first set. What would you tell her?

12. You are a medical examiner collecting bones from a building where there was a very intense fire. You find that many of the bones are brittle and crush easily. **Explain** this observation.

13. **(Circle one answer.)**

    The upper end of the patella is connected to a large muscle in the thigh by a **tendon/ligament.**

    The lower end of the patella is connected to the tibia by a **tendon/ligament.**

# EXERCISE 7

# Examination of Skeletal Structure

## Objectives
After completing this exercise, you should be able to:
- Identify the bones of the axial and appendicular skeletons when articulated and disarticulated
- Locate each bone using correct anatomical terminology
- Differentiate between left and right for paired bones from a disarticulated skeleton
- Determine the gender of an individual from the pelvic bones
- Use the technique of measuring femur length to estimate height
- Apply your knowledge of skeletal analysis to real-life situations.

## CONTENT FOCUS

The human body is composed of 206 bones divided into two groups: the **axial** skeleton and the **appendicular** skeleton.

- **Axial skeleton**—Bones that lie **near the** midline (**axis**) of the body.

- **Appendicular skeleton**—Bones of the **appendages** (such as arms, hips, shoulder, and legs).

These bones provide the framework for support of the human body. By studying the different bones of the skeletal system, scientists are able to gather information about individuals and populations. Patterns of disease, nutrition, migration, mortality, reproduction, and other factors can be explored by analyzing the remains of ancient cultures. Currently, there are several popular television programs that rely on forensic evidence to solve cases. The physical evidence they collect includes information gathered by the examination of bones.

Anthropologists and forensic scientists can use the information gathered from bones to help answer several questions, including:

- Which bones are present? Which are missing?
- Are these bones from one individual or many?
- What was the sex of the individual?
- What was the approximate age of the person at the time of death?
- What was the approximate height of the individual?

The answers to these questions can provide valuable leads to an investigating agency and help identify the remains of an unknown person.

## ACTIVITY 1 — THE HUMAN SKELETON

1. Work **in groups**. Using the information in **Table 7-1**, label all the bones identified on the diagram in **Figure 7-1**. Use the **correct terminology** (not the common names) for each bone you label.

2. Compare your labeled diagram with the articulated skeleton in the classroom. A picture may be worth a thousand words, but there is no substitute for the real thing.

**Exercise 7** ■ *Examination of Skeletal Structure*

## TABLE 7-1
## BONES OF THE ADULT SKELETON

| Axial Skeleton | Appendicular Skeleton |
|---|---|
| skull 22 | **pectoral girdle (shoulder):** |
| hyoid 1 | clavicle (collar bone) 2 |
| auditory ossicles (inner ear) 6 | scapula (shoulder blade) 2 |
| vertebral column (spine) 26 | **upper extremities:** |
| sternum (breastbone) 1 | humerus (upper arm) 2 |
| ribs 24 | ulna (forearm) 2 |
| | radius (forearm) 2 |
| | carpals (wrist) 16 |
| | metacarpals (body of hand) 10 |
| | phalanges (fingers) 28 |
| | **pelvic girdle (hip):** |
| | hip bones 2 |
| | **lower extremities:** |
| | femur (upper leg) 2 |
| | fibula (lower leg) 2 |
| | tibia (shin bone, lower leg) 2 |
| | patella (kneecap) 2 |
| | tarsals (ankle) 14 |
| | metatarsals (body of foot) 10 |
| | phalanges (toes) 28 |
| **Total: 80** | **Total: 26** |

**FIGURE 7-1.** Bones of the Adult Skeleton

 Check your Figure 7-1 labels with your instructor before you continue.

**Exercise 7** ■ *Examination of Skeletal Structure* 121

# ACTIVITY 2                      NAVIGATING THE BODY ANATOMY

Anatomical Position

In the standard anatomical reference position, the body is viewed from the front with the palms facing forward (see **Figure 7-2**). All descriptions of bones are based on this orientation of the human body.

**FIGURE 7-2.** Anatomical Position

Definitions of Anatomical Terms

| TABLE 7-2 DEFINITIONS OF ANATOMICAL TERMS | |
|---|---|
| ANATOMICAL TERM | DEFINITION |
| proximal | close to its attachment to the body (for example, the elbow is proximal to the wrist) |
| distal | away from its attachment to the body (for example, the fingers are distal to the wrist) |
| superior | above (for example, the shoulders are superior to the hips) |
| inferior | below (for example, the mouth is inferior to the nose) |
| medial | closer to the body midline (for example, in the lower arm, the ulna is medial to the radius) |
| lateral | away from the body midline (for example, the ear is lateral to the nose) |
| anterior | front side of the body (for example, the chest is anterior to the buttocks) |
| posterior | back side of the body (for example, the spine is posterior to the stomach) |

**Note:**

In humans, the terms **anterior and ventral** both refer to the front side of the body (the side that faces forward when we walk). In four legged animals (such as the fetal pig), however, the front side of the body faces the ground, and thus is referred to as **ventral**, while the term anterior refers to the head end of the animal.

Using the information in **Table 7-2**, fill in the blanks with the correct anatomical terms:

1. The sternum is _____ to the heart.

2. The skull is _____ to the hip bones.

3. The nose is _____ to the ears.

4. The shoulder is _____ to the elbow.

**Exercise 7** ■ *Examination of Skeletal Structure*

5. The tarsals are _____ to the tibia.

6. The patella is _____ to the clavicle.

7. The vertebral column is _____ to the sternum.

8. The index finger is _____ to the ring finger.

☐ **Check your answers with your instructor before you continue.**

## ACTIVITY 3   TELLING YOUR LEFT FROM YOUR RIGHT

*Preparation*

It's easy to differentiate between the left and right sides of the skeleton when it is articulated. The problem, however, is that bones are often found **separated** from the rest of the skeleton. How can these **disarticulated (single)** bones be identified?

Each bone has certain characteristics that distinguish it from other similar bones. The appearance of various bone features and their locations on a particular bone are used by forensic anthropologists to determine if the bone came from the right or left side of the body.

1. Work in groups. From the supply area, obtain a **numbered container of disarticulated bones**.

2. Use the following information to **identify each bone** in your container.

> **Note:**
> Don't panic at the technical names for bone features. They are only to be used for identification and side determination. **DO NOT** memorize the names of all these features without consulting your instructor. Your instructor will help you determine which information is most important.

3. **If possible, determine whether each of the following bones came from the right or left side of the body**: humerus, femur, and patella.

   **Side determination of a complete humerus (see Figure 7-3):**
   - Hold the bone with the head up. Look at the opposite end of the bone. Notice that both sides have an indentation. The indentation that is deeper and larger in size is on the posterior surface. Place the bone over your own right or left humerus such that **posterior surface is touching your arm and the head is medial**. You have now identified the right or left humerus.

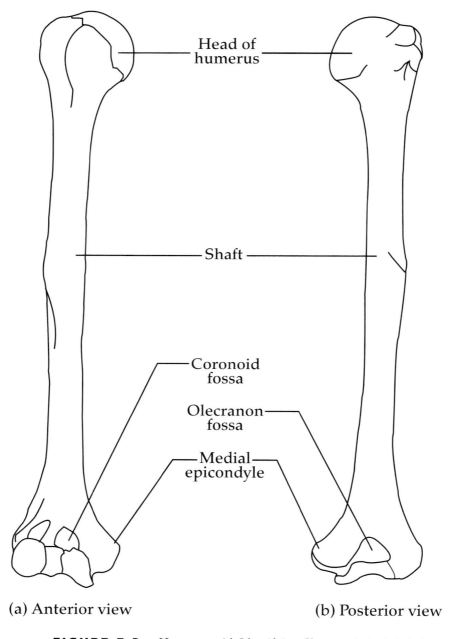

(a) Anterior view    (b) Posterior view

**FIGURE 7-3.** Humerus with Identifying Characteristics Labeled

**Exercise 7** ■ *Examination of Skeletal Structure*

### Side determination of a complete femur (see Figure 7-4):

- Hold the bone with the head up. Look at the opposite end of the bone. Notice that both sides have an indentation. The indentation that is deeper and larger in size is on the posterior surface. Place the bone over your own right or left femur such that **posterior surface is touching your thigh and the head is medial**. You have now identified the right or left femur.

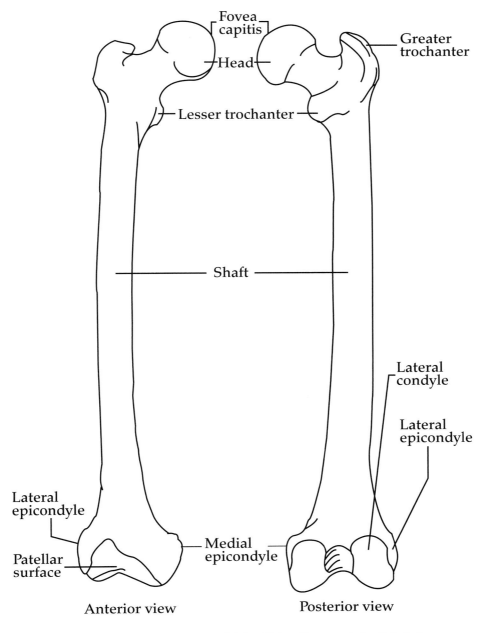

**FIGURE 7-4.** Femur with Identifying Characteristics Labeled

**Side determination of a complete patella (see Figure 7-5):**

- Hold the bone with the posterior side facing you and the **apex** (pointed end) down. Note the two depressions (called **articular facets**) that mark the site of articulation with the base of the femur. **The larger (lateral) articular facet is on the side that the bone is from**.

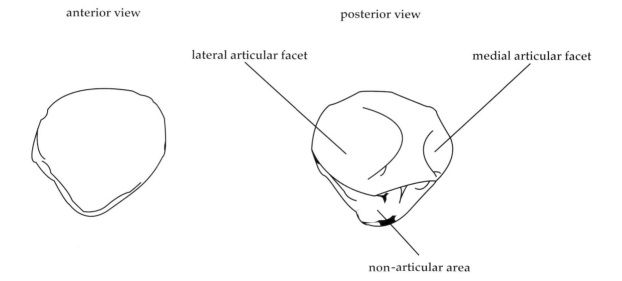

**FIGURE 7-5.** Patella with Identifying Characteristics Labeled

**Exercise 7** ■ *Examination of Skeletal Structure*

4. Record the results of your side determinations for the humerus, femur, and patella in **Table 7-3**.

| TABLE 7-3 SIDE DETERMINATION OF DISARTICULATED HUMAN BONES ||
|---|---|
| Container ID#: | |
| Name of Bone | Side Determination (Left or Right) |
|  |  |
|  |  |
|  |  |
|  |  |
|  |  |

**Check your answers with your instructor before you continue.**

✓ Comprehension Check

1. In a recent excavation of the city of Alexandria, scientists found a large collection of disarticulated human bones: however, there were no skulls. The inventory of the find is as follows:

   | | |
   |---|---|
   | individual vertebrae | 62 |
   | clavicle | 2 |
   | ribs | 24 |
   | radius | 3 |
   | sternum | 1 |
   | femur | 5 |
   | fibula | 3 |
   | tibia | 6 |
   | patella | 1 |

   Based on the inventory of bones discovered, what is the **minimum number of people** that could be represented in this collection? **Explain** your answer in detail.

## ACTIVITY 4 — DETERMINING GENDER FROM THE PELVIC BONES

Preparation

You have probably noticed that women, generally, have broader hips than men. This characteristic is due to differences between the male and female pelvis.

The pelvis is composed of a pair of irregularly shaped bones called the **hip bones**. Each adult hip bone is formed by the fusion of three separate bones, **the ilium, the ischium, and the pubis**, which unite during adolescence (see **Figure 7-6**).

a. **Ilium—Upper** portion of the pelvis.

b. **Ischium—Lower** portion of the pelvis (supports the sitting body).

c. **Pubis—Anterior** portion of the pelvis. The anterior articulation of the hip bones is called the **pubic symphysis**.

**Exercise 7** ■ *Examination of Skeletal Structure*

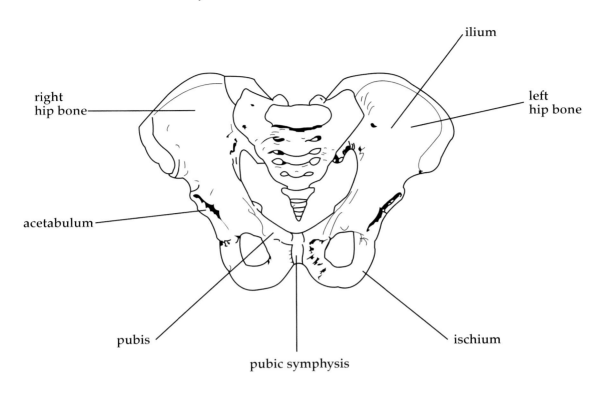

**FIGURE 7-6.** Bones of the Pelvis

**Gender determination (see Figure 7-7):**

1. Work in groups. Select a pelvis from your box of disarticulated bones.

2. In the **male hip bones**, the pubis is shorter, causing the subpubic angle to approximately resemble the **letter "V"** in an articulated pelvis.

   ■ Place your index finger flat against the pubic symphysis, allowing your thumb to rest in the subpubic space. If your thumb can move only slightly or not at all, the sex is probably **male**.

   ■ In the **female hip bones**, the pubis is longer, causing the subpubic angle to approximately resemble the **letter "U"** in an articulated pelvis.

   ■ If your thumb has more room for movement, this is an indication that the hip bones bone belongs to a **female**.

   ■ In general, the pelvis of a male is narrower, heavier and rougher than in females (for muscle attachment). The pelvic girdle of a female is wider, lighter, and smoother.

   ■ The **pelvic opening** is generally **large and oval in females** and **small and heart shaped in males**.

**FIGURE 7-7.** The Male and Female Pelvis

**Exercise 7** ▪ *Examination of Skeletal Structure* **131**

3. The **acetabulum**, the articulation site for the head of the femur, is **larger in males than females**. What does this tell you about the difference between the **head of the femur** in males and females?

**Gender determination guides using other bones of the body:**

- Generally, male bones are larger, heavier, and rougher than female bones.

- The female skull is generally smaller and rounder than the male. It exhibits a vertically longer forehead and a smaller jaw.

- The female sacrum is broad and short compared to the longer, narrower male sacrum.

- The female coccyx (tailbone) is more moveable than the male coccyx.

- Several of the bones of the body can be measured and compared to reference standards for gender, height, and age estimation.

## Comprehension Check

1. You are an expert on skeletal remains. You receive a phone call from a local police department, which has recovered a pelvic bone that was found in the woods by a hiker. They would prefer not to waste time sending the bone through the mail, and ask you to describe to them, over the phone, the method they should use to determine if the bone came from a male or female. **Be clear and complete**.

2. Some bones were recovered from two burials in a cemetery with unmarked graves. Both skeletons had all the bones present. In the skeleton #1, the subpubic angle was clearly "V" shaped and the sacrum was long and narrow. In skeleton #2, the subpubic angle was clearly "U" shaped with a broad sacrum.

What was the gender of each individual?

If the pelvic bones were missing from the skeleton, could the number of ribs be used to determine the gender of each individual instead? **Explain your answer.**

**Check your answers with your instructor before you continue.**

## ACTIVITY 5 — ESTIMATING BODY HEIGHT FROM FEMUR LENGTH

*Preparation*

An individual's height can be estimated from skeletal remains even if the skeleton is disarticulated. There are several different ways by which height can be estimated. In all cases, there is a margin of error (expressed as +/− in the calculated height of the individual), which must be taken into consideration when trying to identify the skeletal remains.

In this activity, you will be making height estimates based on femur length.

1. Work in groups. Select a femur from your box of disarticulated bones. Get the following: a **tape measure**.

2. Measure the diameter of the head of the femur, in **millimeters (mm)**. Record your measurement: _____ **mm**

**Exercise 7** ■ *Examination of Skeletal Structure*

3. Measure the mid-shaft circumference of the femur in **millimeters (mm)**. Record your measurement: _____ **mm**

4. Measure the length of the femur from the **greater trochanter to the lateral condyle** in **centimeters (cm)**. Record your measurement: _____ **cm**

5. Compare your femur measurements with the information in **Table 7-4** to determine the gender of the individual from which the femur came.

### TABLE 7-4
### DETERMINATION OF GENDER FROM FEMUR MEASUREMENTS

| MEASUREMENTS | FEMALE | GENDER DETERMINATION QUESTIONABLE | MALE |
| --- | --- | --- | --- |
| Diameter of Head of Femur | < 43 mm | 43–47 | > 47 mm |
| Mid-shaft Circumference of Femur | < 81 mm | 81 mm | > 81 mm |
| Femur Length | < 39 cm | 40–43 cm | > 43 |

**Note:**
If all the above femur measurements fall into the "Gender Determination Questionable" range, choose another femur and measure again.

6. What was the probable gender of the individual that this femur belonged to?
_____

Did **both the femur and the pelvis in your box of disarticulated bones** come from the same gender individual? **Explain your answer.**

7. **Based on your gender determination,** use the formula below and your femur length measurement to calculate the estimated height of the individual.

   for **male** skeletons:

   **estimated height = (measured length of femur × 2.32) + 65.53**

   for **female** skeletons:

   **estimated height = (measured length of femur × 2.47) + 54.10**

> **Note:**
> For the male femurs, the margin of error in the femur length calculation is +/− 6.96 cm. In females, the margin of error in the femur length calculation is +/− 5.96 cm.

8. Based on your femur calculations, how tall was the individuals the femur came from? in cm _____ converted to feet and inches _____

   **(Hint: One inch = 2.54 centimeters.)**

9. Explain how information on gender and height can be used to help identify a set of skeletal remains.

## Comprehension Check

1. A construction company is excavating a site for a new building. In the process of digging for the foundation, a few bones are discovered. Old stories say that a woman disappeared from a house that formerly stood on the site, and her ghost haunts the premises.

   Among the bones was a femur with a head diameter of 46 mm and a femur length of 50 cm. Have the construction workers found her remains? **Explain your answer**.

**Exercise 7** ■ *Examination of Skeletal Structure* 135

Self Test

Enter the letter of the correct bone for each of the following descriptions. Answers can be used **more than once** and questions may have **more than one** correct answer.

a. clavicle
b. carpals
c. femur
d. fibula
e. humerus
f. hyoid bone
g. hip bones
h. metacarpals
i. metatarsals
j. phalanges
k. patella
l. radius
m. ribs
n. scapula
o. sternum
p. tarsals
q. tibia
r. ulna
s. vertebrae

1. \_\_\_\_\_ Small bone that comes in contact with the ground when you kneel
2. \_\_\_\_\_ Lateral bone of the leg
3. \_\_\_\_\_ Bone that is medial and anterior to the ribs
4. \_\_\_\_\_ Bones forming the knee joint
5. \_\_\_\_\_ Bone affected if you bumped the anterior part of your lower leg
6. \_\_\_\_\_ Shoulder blade
7. \_\_\_\_\_ Collar bone
8. \_\_\_\_\_ Breastbone
9. \_\_\_\_\_ Medial bone of the leg
10. \_\_\_\_\_ Bone that is distal to the clavicle
11. \_\_\_\_\_ Bones of the ankle
12. \_\_\_\_\_ Bones of fingers and toes
13. \_\_\_\_\_ Medial bone of the forearm
14. \_\_\_\_\_ Lateral bone of the forearm
15. \_\_\_\_\_ Bones of the pelvis
16. \_\_\_\_\_ Bones encompassed by your wristwatch
17. \_\_\_\_\_ Bones inside your tennis shoe

18. In an articulated skeleton, the ribs are connected to the sternum with cartilage. An artificial material replaces this cartilage in the classroom skeleton.

    Explain why this design functions better than having the bones of the rib cage connected directly to the sternum.

# EXERCISE 8

# The Nervous System

## Objectives
After completing this exercise, you should be able to:
- identify and explain the function of each of the major structures of the brain and spinal cord
- explain the effects of damage to various regions of the brain and spinal cord
- discuss the differences in function among sensory, motor, and interneurons
- diagram and explain the sequence of actions in a spinal reflex arc
- compare and contrast grey matter and white matter
- draw conclusions that are supported by experimental data
- analyze data using common statistical measures
- apply your knowledge of the scientific method to real-life situations.

## CONTENT FOCUS

To function properly, all of your organ systems must coordinate and communicate with the nervous system. Every activity in your daily life is made possible through the action of the nervous system. Since the nervous system plays such an integral role in your life, it is important to have a working knowledge of basic nervous system functions.

The **nervous system** responds **rapidly** to environmental stimuli. This rapid response time allows you to cross a busy street, react to emergency situations, read this book, or use a computer keyboard.

In this exercise, you will be examining the structure of the **central nervous system** (the **brain and spinal cord**). You will also have a chance to explore various functions that are performed by these organs.

**FIGURE 8-1.** Longitudinal View of the Human Skull and Brain

## ACTIVITY 1     STRUCTURE AND FUNCTION OF THE BRAIN

Preparation

To complete this activity, refer to **Figures 8-1 and 8-2**. You may also refer to human brain models, preserved brains, or other materials present in the classroom.

1. The brain is one of the body's most vital organs, and as such, needs protection. The first protective barrier consists of the bones of the skull. Although the skull does a good job of protecting the brain from external impact, internal cushioning is also required, since even a slight impact can cause the brain to hit the inside of the skull and result in a serious injury. To prevent this, the brain is cushioned and protected by three layers of connective tissue called the **meninges** (see **Figure 8-1**). Additional cushioning is provided by cerebrospinal **fluid between the layers**.

Medical problems related to the brain and meninges are often mentioned in the news media. For example, **meningitis** is an infection of the cerebrospinal fluid surrounding the brain or spinal cord. Meningitis is usually caused by a viral or bacterial infection that can damage or kill neurons and cause serious health problems.

Exercise 8 ■ *The Nervous System*

Center for abstract thinking and problem solving, controls social behavior and expression of emotions. Controls voluntary motor activities and memories.

**Frontal Lobe**

Interprets sensations of touch, pressure, temperature, and taste. Coordinates information from a variety of senses.

**Parietal Lobe**

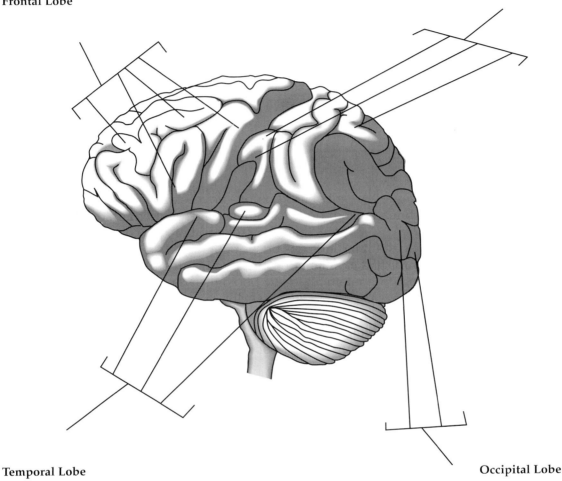

**Temporal Lobe**

Interprets and forms memories from auditory experiences. Functions in hearing, language skills, and understanding speech.

**Occipital Lobe**

Interprets visual images and coordinates vision with information from the other senses.

**FIGURE 8-2.**

2. The **cerebrum** (see **Figure 8-1**) is the largest part of the brain, covering almost the entire superior surface. It has two halves (called **hemispheres**), each controlling the **opposite side of the body**. The two halves are connected by a broad band of nerve tissues called the **corpus callosum**.

What part of the body is affected if the **right side of the brain is injured** in an accident?

Observe the cerebrum and note its compact, folded structure. Within the body there are many different examples of ways to increase surface area and maximize functional capabilities. One of the most common ways is to **layer or fold** the tissue to fit into a smaller space. The cerebrum is an excellent example of this principle, since its highly folded surface is one of the most noticeable features of this part of the brain.

Each cerebral hemisphere is **divided into regions called lobes**, which control various aspects of conscious thought and problem solving abilities (see **Figure 8-2**).

3. Locate the **cerebellum**, a large, rounded structure positioned just below the occipital lobes of the cerebrum (see **Figures 8-1 and 8-2**).

   The cerebellum processes information from the senses, joints, and muscles to coordinate movement of skeletal muscles, posture, balance, and equilibrium.

## Comprehension Check

Based on the diagram in **Figure 8-2**, fill in the blanks with the choice that is **most appropriate** to locate the part of the cerebrum that is most likely to be malfunctioning. Answers can be used **more than once**. Some questions may have **more than one correct answer**.

a. Frontal lobe
b. Temporal lobe
c. Parietal lobe
d. Occipital lobe

1. _____ Difficulty identifying colors.

2. _____ Difficulty understanding spoken words.

3. _____ Can't thread a needle.

4. _____ Doesn't feel pain when pricked with a pin.

5. _____ Noticeable changes in personality.

# Exercise 8 ■ *The Nervous System*

6. Since the cerebellum is related to coordinating movements, do you think the relative size of the cerebellum would be **larger or smaller in birds than in humans**? **Explain** your answer.

**Check your answers with your instructor before you continue.**

4. The **brain stem** (see **Figure 8-1**) connects the brain to the spinal cord. It is composed of three parts: the **midbrain**, the **pons**, and the **medulla oblongata**.

   All of the sensory and motor neurons carrying information to other areas of the brain pass through the brain stem. The brain stem sorts and directs information to the appropriate brain centers and acts as the main "highway" for nerve tracts traveling between the brain and spinal cord.

   **Nuclei** (localized control centers) within the **medulla oblongata** regulate several automatic body functions, such as heartbeat, breathing, and blood pressure.

5. There are four **ventricles** within the brain (see **Figure 8-1**). The ventricles form a series of **connected chambers** within the cerebrum and brain stem. The ventricles are filled with **cerebrospinal fluid** that circulates throughout the ventricles and the central canal of the spinal cord. The fluid circulates between the brain and spinal cord in the space between the meninges where it acts as a shock absorber, provides nutrients, and removes wastes.

6. Several additional internal structures can be identified on **Figure 8-1**. These include the thalamus, the hypothalamus, and the pituitary gland.

   The **thalamus** contains about a dozen specialized nuclei that sort and group nerve impulses by function and relay them to the appropriate regions of the cerebrum for action.

The **hypothalamus** is one of the body's most important control centers for **homeostasis** (regulation of the internal environment). Among the important factors controlled by the hypothalamus are regulation of body temperature, hunger and thirst, blood sugar level, water balance, and emotional responses (such as sex drive).

As part of its homeostatic function, the hypothalamus monitors body conditions, and responds by releasing **regulating factors**. Regulating factors are chemicals that stimulate or inhibit the release of hormones from the pituitary gland.

The hypothalamus is linked to the **pituitary gland** (an endocrine gland located just beneath the hypothalamus). The pituitary gland produces a number of hormones that help to regulate growth, kidney function, reproductive cycles, milk production, and basal metabolic rate.

### Comprehension Check

1. Name **at least two** parts of the brain that would be activated when **riding a bicycle**. **Explain** your answer.

2. Name **five** parts of the brain that could be activated when you see an attractive person passing by in a string bikini.

## ACTIVITY 2   SPINAL NERVES AND THE REFLEX ARC

*Preparation*

The nervous system has several different types of nerve cells (**neurons**). **Sensory neurons** process and transmit incoming information to the central nervous system (the brain and spinal cord). **Motor neurons** carry information to muscles or glands, and the muscles or glands respond with the appropriate actions. **Interneurons** form **connections** between neurons.

**Exercise 8** ■ *The Nervous System*

**Reflexes** are unconscious, **involuntary motor responses**. A familiar spinal reflex occurs when you accidentally touch a hot stove. Before you even realize that your hand is painfully hot, a reflex has jerked your hand back from the stove.

During a **spinal reflex**, a sensory receptor is stimulated and the impulse is transmitted to the spinal cord. Many spinal reflexes are mediated only by the spinal cord and do not involve higher brain centers. This means that a motor neuron response can be stimulated by the spinal cord alone without input from the brain, although the brain can influence the strength of reflex responses.

Neurons vary in function and appearance, but all share a similar three-part structure: the cell body, the axon, and the dendrites. Every neuron has a **cell body**, the region of the cell that contains the nucleus and a variety of other organelles necessary for cell metabolism. The cell bodies of **most** neurons are located **within the central nervous system**; however, some are located **outside the CNS** in clusters called **ganglia**.

Outside stimuli are received by **dendrites** in the skin and other sense organs. Incoming signals from these dendrites are transmitted toward the cell body and continue into the **axon**. Axons relay outgoing messages from one neuron to other neurons or to various tissues and organs in the body.

1. Refer to the diagram in **Figure 8-3**.

   Label the following: **sensory neuron, interneuron, motor neuron, and the cell bodies of each of the three neurons**.

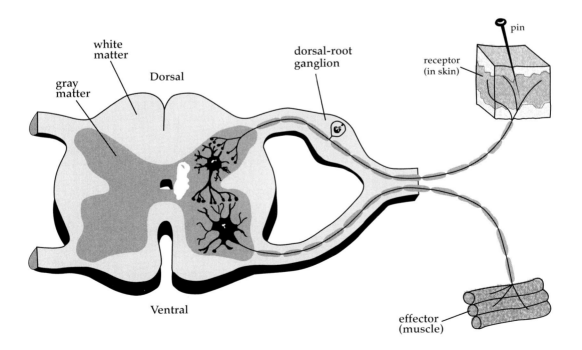

**FIGURE 8-3.** Reflex Arc in the Spinal Cord

2. In which direction do impulses travel through sensory neurons?

   In which direction do impulses travel through motor neurons?

   **(Circle one answer.) Sensory neurons / interneurons / motor neurons** have the longest axons.

3. Work in **groups of two students**. Get the following supplies: **two rubber reflex hammers**.

   Complete the following two activities to demonstrate the knee-jerk and plantar reflexes with your partner, then form a group of four students to complete the third reflex activity.

**Knee-Jerk Reflex**

1. Sit on a laboratory stool or table with your **legs relaxed and hanging freely. Make sure your feet do not touch the floor.**

2. With your finger, locate a soft area just below the patella (the knee cap).

   Gently tap this area with the reflex hammer and observe the response.

   > **Note:**
   > **It make take you several tries to locate the exact spot to produce the patellar reflex response.**

3. What were the results of your experiment?

4. **List, in order**, the events that occur in the nervous system during the knee-jerk reflex. Begin when the hammer hits the knee and conclude with the knee-jerk response.

**Exercise 8** ■ *The Nervous System*

5. **Exchange roles** and complete the Plantar Reflex activity.

**Plantar Reflex (Negative Babinski Reflex)**

1. Remove the shoe and sock from one foot.

   Sit on a laboratory stool and **support your foot** at the ankle on a second stool.

2. Gently draw the metal tip of the handle of the reflex hammer along the side of the sole **from the heel toward the big toe (across the ball of the foot)**. Observe the response.

**Blink Reflex**

1. Work in **groups of four students**. Get the following supplies: **clear plastic wrap and a piece of cotton**.

   Form the cotton into a large ball, about the size of a golf ball.

   Assign one member of your group to be the first test subject. A second group member will conduct the experiment, and the remaining members will observe and record the results of each trial.

   The test subject will hold the piece of plastic wrap in front of his/her face and eyes, far enough away from the face to serve as a barrier. The test subject should be **relaxed**, but attentive.

2. Throw the cotton ball at the plastic wrap. In **Table 8-1**, record if the test subject blinked or did not blink. **Repeat** the experiment **ten** times.

   **Exchange roles and repeat the experiment** until **each group member** has had an opportunity to be the test subject.

3. What percentage of trials produced blink responses for each test subject?

   **Subject 1** _____ %

   **Subject 2** _____ %

   **Subject 3** _____ %

   **Subject 4** _____ %

| TABLE 8-1 | | | | |
|---|---|---|---|---|
| **BLINK REFLEX RESULTS** | | | | |
| Trials | Blink (Yes/No) Subject 1 | Blink (Yes/No) Subject 2 | Blink (Yes/No) Subject 3 | Blink (Yes/No) Subject 4 |
| 1 | | | | |
| 2 | | | | |
| 3 | | | | |
| 4 | | | | |
| 5 | | | | |
| 6 | | | | |
| 7 | | | | |
| 8 | | | | |
| 9 | | | | |
| 10 | | | | |
| TOTAL (Number of blinks) | | | | |

4. Since all the test subjects knew that the plastic wrap would prevent the cotton ball from hitting their faces, why did the blink response occur? **Explain** your answer.

**Exercise 8** ■ *The Nervous System*   **147**

### ✓ Comprehension Check

You're walking barefoot along the boardwalk at the beach, gawking at members of the opposite sex, when all of a sudden, you step on a sharp object. The following activities will be occurring in your nervous system. Place the actions in the correct sequence.

1. _____ Muscles in your leg contract

2. _____ Sensory nerve endings are stimulated

3. _____ Brain senses pain

4. _____ Sensory neurons transmit impulses to the spinal cord

5. _____ Motor neurons transmit impulses to leg muscles

6. _____ Foot moves

7. _____ Members of opposite sex no longer seem so important

**Check your answers with your instructor before you continue.**

## ACTIVITY 3 — STRUCTURE AND FUNCTION OF THE SPINAL CORD

1. Get the following supplies: **a compound microscope and a prepared slide labeled "cross section of the spinal cord."**

   View the slide using the **scanning lens (4X)** of your microscope.

2. You will note a "butterfly" shaped area with a darker stain towards the center of the spinal cord. This area of the spinal cord is composed **cell bodies** of interneurons and motor neurons. The tissue is known as **grey matter** because it appears grey to the naked eye.

   The white matter is composed of nerve fibers covered by a **myelin sheath**, which is wrapped around the axons of many neurons. When a signal travels along a myelinated axon, it can travel more than 300 miles per hour. Signals travel less than five miles per hour in axons without myelin sheaths.

3. **Spinal nerves**, which contain **nerve fibers** (bundled axons and dendrites), lie along the length of the spinal cord, but they can be difficult to see. Examples of spinal nerves can be seen in **Figure 8-3**.

   Refer to the figure. You can see that the spinal nerves divide into two pathways just prior to entering the spinal cord. These are called the **dorsal and ventral roots**. The **dorsal root** of a spinal nerve contains **sensory fibers** entering the gray matter of the spinal cord, and the **ventral root** of a spinal nerve contains **motor fibers** leaving the gray matter.

4. Focus your microscope on an area within the grey matter and switch up to **high power (40X)**.

   Locate one or more neurons and observe the location of the following cell structures: **cell body, nucleus, cell membrane, axon, and dendrites**.

## ✓ Comprehension Check

Based on your microscopic observations and the diagram in **Figure 8-3**, answer the following questions.

1. **(Circle one answer.)**

   If an injury affected **only the dorsal root** of a spinal nerve, **sensory input / motor output** would be primarily affected.

   The cell body of the motor neuron is located in the **gray matter / white matter**.

   The dendrites of a motor neuron are located in the **grey matter / white matter / dorsal root / ventral root**.

   The cell body of a sensory neuron is located in the **grey matter / white matter / dorsal root / ventral root**.

2. **(Circle all that apply.)**

   The nerve fibers of **sensory neurons / motor neurons / interneurons** are located in spinal nerves.

   The myelinated axons of motor neurons are located in the **grey matter / white matter / dorsal root / ventral root**.

# Exercise 8 ■ *The Nervous System*

3. In anatomy, spinal nerves are sometimes classified as "**mixed nerves**." Why would this description be appropriate?

**Check your answers with your instructor before you continue.**

## ACTIVITY 4                                                           REACTION TIME

*Preparation*

A reaction is a **conscious, voluntary response** to an outside event. As discussed in Activity 1, voluntary responses to stimuli are mediated by the cerebral cortex. You exert conscious control of your muscles when you chew food, manipulate your knife and fork, or take notes in class.

During a voluntary response, sensory receptors are stimulated and the impulse is transmitted through the spinal cord to higher brain centers. The incoming signal and response travel a greater distance than in a spinal reflex (which is mediated by the spinal cord), therefore, a reaction response requires more time to complete than a reflex.

1. Work **with a partner**. Get the following supplies: **a reaction time ruler.**

2. Sit on a laboratory stool. You are the test subject. Your partner will stand next to you. He or she is the investigator conducting the reaction time experiment.

   The investigator will hold the meter stick vertically with the **release end** toward the **ceiling**, so that the **thumb line** of the ruler is approximately **chest height** on the seated **test subject**.

The test subject should position the **thumb and fingers of one hand** around the meter stick at the location labeled **thumb line**. The thumb and fingers must be about **one inch away** from the sides of the ruler.

3. Once the investigator and test subject are correctly positioned, you can begin the experiment.

   **Without warning the test subject**, the investigator will release the meter stick and the test subject will catch the meter stick as quickly as possible.

   **Do not release your grip on the meter stick!**

   The investigator will observe the number on the stick closest to the **position of the test subject's thumb**. The ruler is calibrated such that the **position of the thumb** represents the test subject's **reaction time in milliseconds**. Record the results in **Table 8-2**.

4. Repeat the procedure nine more times, recording the results of each trial in **Table 8-2**.

   If time permits, switch places and repeat the experiment. **Calculate** the **mean** reaction time for each test subject.

| TABLE 8-2 REACTION TIME RESULTS | | |
|---|---|---|
| TRIALS | SUBJECT 1 RESPONSE TIME (mSEC) | SUBJECT 2 RESPONSE TIME (mSEC) |
| 1 | | |
| 2 | | |
| 3 | | |
| 4 | | |
| 5 | | |
| 6 | | |
| 7 | | |
| 8 | | |
| 9 | | |
| 10 | | |
| Mean Reaction Time | | |

EXERCISE 9

# Introduction to Forensic Biology

## Objectives
After completing this exercise, you should be able to:
- identify and interpret basic fingerprint patterns
- use your knowledge of fingerprint patterns and details to analyze forensic evidence
- demonstrate an understanding of the A-B-O and Rh blood typing systems
- use antigen–antibody interactions to identify and type unknown blood samples
- apply your knowledge of basic forensic techniques to real-life situations.

## CONTENT FOCUS

A scientist is similar to a detective. Both use scientific thinking to solve problems without clear solutions. Solving this type of open-ended problem requires you to develop your skills in forming hypotheses, gathering evidence by observation and experimentation, and drawing conclusions on the basis of your findings.

Forensic science is a good example of the practical application of the scientific method—in this case, science is used to analyze evidence and solve crimes. Forensic scientists use information from all branches of science (including chemistry, geology, physics, medicine, and, of course, **biology**).

This exercise provides an introduction to two of the basic techniques that are employed in forensic biology: **fingerprint analysis and blood typing**.

## ACTIVITY 1  GETTING STARTED

*Preparation*

**Fingerprinting** is the most commonly used and reliable system of identification. Each person has a unique pattern of ridges and valleys on their fingertips. Although other physical characteristics (such as weight, height, and hair color) may change as a person ages, fingerprints remain the same for life. Fingerprints penetrate through five skin layers and so are almost impossible to erase. Criminals have tried filing, acid burning, and even surgical removal, but with limited success.

Fingerprints have been used for identification purposes for several thousand years. The Babylonians recorded their fingerprints in the soft clay of their writing tablets (paper was not yet in use). The fingerprints served as a "signature" to prevent document forgeries. A similar system was also used in ancient China and Japan. In the last 100 years, an internationally recognized system of identifying fingerprints has been developed. Most law enforcement agencies (even in local communities) maintain their own fingerprint files. They can also access fingerprint information from state agencies and the FBI.

The skin is covered with **sweat glands**, including some in the ridges on the fingers and palms. During normal activities, perspiration from the sweat glands accumulates in these ridges. These ridges also accumulate body oils (from touching oil-producing areas such as the face, scalp, or neck). When you touch an object, perspiration and skin oils are transferred to that surface, leaving an invisible impression (the **latent fingerprint**). Latent prints found on nonabsorbent surfaces (such as metal or glass) can be dusted with colored powder and removed with transparent tape. Latent prints on absorbent surfaces (such as cloth or paper) must be recovered chemically.

The **Fingerprint Classification System** has three basic patterns: **arches, loops, and whorls**.

1. Begin by learning to recognize the basic identifying patterns used in fingerprint analysis.

    | | |
    |---|---|
    | **Pattern area** | Part of the fingerprint that is used for identification (**Figure 9-1a**) |
    | **Ridgelines** | Individual lines that make up a fingerprint |
    | **Delta** | Area where **two ridgelines** come to a **point** (**Figure 9-1b**) |
    | **Core** | Approximate center of the finger impression (**Figure 9-1b**) |
    | **Whorl** | **Complete circle** formed by ridgelines between two deltas; whorls **do not** continue off either side of the print (**Figure 9-2**) |

# Exercise 9 ■ Introduction to Forensic Biology

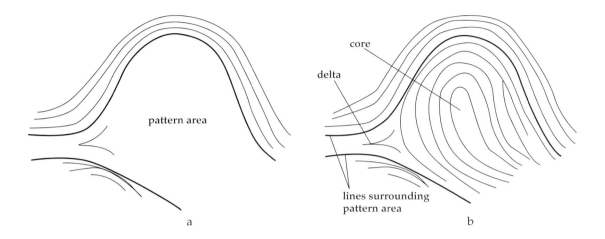

**FIGURE 9-1.** (a) Fingerprint Pattern Area and (b) Features Used in Identifying Fingerprints

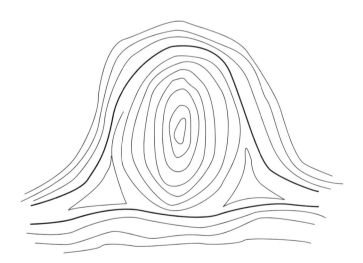

**FIGURE 9-2.** Fingerprint with Whorl

## ✓ Comprehension Check

1. Refer to **Figure 9-1b**. How many **ridgelines** are between the delta and the core?
   _____

2. **(Circle one answer.)** The fingerprint in **Figure 9-1b** is/is not a whorl.

3. Refer to **Figure 9-2**. How many **ridgelines** are between the delta and the core?
   _____

2. In addition to whorls, there are several other common fingerprint patterns.

   | | |
   |---|---|
   | **Simple arch** | Ridgelines **enter** on the **left**, rise, and **exit** on the **right** of the print (**Figure 9-3a**) |
   | **Tented arch** | Ridges meet at the center and form a **peak** (**Figure 9-3b**) |

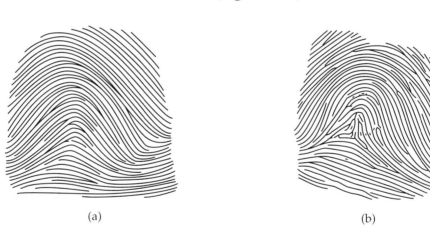

(a)  (b)

**FIGURE 9-3.** (a) Simple Arch and (b) Tented Arch

3. There are various types of loops.

   | | |
   |---|---|
   | **Loops** | Curved ridgelines that **enter** and **exit** on the **same side** of the print (**Figure 9-4a**) |
   | **Double loop** | Two separate loop formations on the same finger, (**Figure 9-4b**) |

(a)                        (b)

**FIGURE 9-4.** (a) Loop and (b) Double Loop

4. Fingerprints that do not fit easily into any of the three categories are sometimes separated into a fourth category, called **mixed**. Mixed prints usually show a combination of two or more different patterns.

5. If additional identifying characteristics are required to make a fingerprint match, the following can be considered:

   | | |
   |---|---|
   | **Bifurcation** | Forking or dividing of one ridgeline into two or more branches **(Figure 9-5)** |
   | **Divergence** | Spreading apart of two ridgelines that have been running parallel or nearly parallel **(Figure 9-5)** |

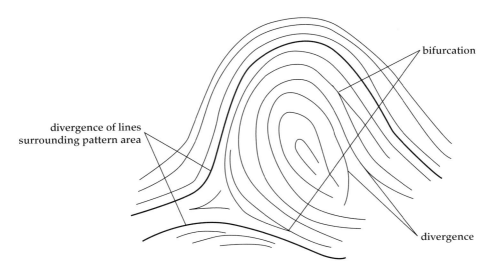

**FIGURE 9-5.** Additional Identifying Features

## Comprehension Check

1. Place an "X" in front of each feature that is found in the fingerprints in **Figure 9-6**.

Print #1                Print #2

| X | Characteristic |
|---|---|
|   | Delta |
|   | Whorl |
|   | Simple arch |
|   | Tented arch |
|   | Loop |
|   | Double loop |
|   | Mixed |
|   | Divergence of ridgelines |
|   | Bifurcated ridgelines |

| X | Characteristic |
|---|---|
|   | Delta |
|   | Whorl |
|   | Simple arch |
|   | Tented arch |
|   | Loop |
|   | Double loop |
|   | Mixed |
|   | Divergence of ridgelines |
|   | Bifurcated ridgelines |

**FIGURE 9-6.** Analysis of Fingerprints

**Check your answers with your instructor before you continue.**

Exercise 9 ■ Introduction to Forensic Biology

# ACTIVITY 2                                    THE FINGERPRINT "FORMULA"

Preparation

A fingerprint formula is a list of **print classifications for each hand**. Each hand has a distinctive fingerprint formula. Before you can analyze crime-scene prints, you must develop some expertise in recognizing the different fingerprint patterns.

1. Work **in groups**. Get the following supplies: **one fingerprint inkpad per group, a magnifying glass, and some paper towels.**

2. **Remove Figure 9-7** from your notebook and place it on the table in front of you. Using the fingerprint inkpad, make a clean set of fingerprints for each member of your team.

   Place the prints on the **Fingerprint File Card** in **Figure 9-7**. To make the prints, place **one side** of each finger on the inkpad. **ROLL** the finger from one side to the other, covering the **FRONT** (**NOT** the tip) with ink.

> **Note:**
> **Be careful with the fingerprint ink. It can stain your clothes.**

3. **LIGHTLY** repeat the rolling motion to transfer the inked print onto **Figure 9-7** (or a clean sheet of white paper).

   **If you press too hard**, you will get a black ink blob with no discernable ridges. Use a very light touch to obtain a readable print.

Left Hand                                                                                    Thumb

|  |  |  |  |  |
|--|--|--|--|--|
|  |  |  |  |  |

Right Hand

Thumb

|  |  |  |  |  |
|--|--|--|--|--|
|  |  |  |  |  |

**FIGURE 9-7.** Fingerprint File Card

4. To prepare an individual's fingerprint formula, **begin at the thumb and proceed to the little finger.** For example, a hand that has fingers of arch, arch, whorl, loop, whorl has a print formula of **a-a-w-l-w**.

   Using your own fingerprints, **determine your fingerprint formula. Write the fingerprint formula directly underneath each fingerprint in Figure 9-7.**

5. **By group discussion, double-check** each person's assessment of their own fingerprint formula.

   Hard-to-classify prints may require your group to clarify and add information to the definitions of each category. As the distinctions between print patterns become more clearly defined, it will be easier for you to assign individual prints to a print type.

 **Check your answers with your instructor before you continue.**

6. Using your revised print type descriptions, **classify the 12 fingerprints in Figure 9-8.**

**Exercise 9** ■ *Introduction to Forensic Biology* 161

**FIGURE 9-8.** Practice Fingerprints

Check your answers with your instructor before you continue.

## ACTIVITY 3 — DETERMINING POINTS OF SIMILARITY

As you can see from looking at the fingerprints in **Figure 9-8**, there are only a few possible fingerprint patterns. Consequently, it is necessary to look at the **fine details of the ridgelines** within the prints (such as bifurcations, divergences, etc.) to match sample fingerprints to a specific person.

**Exact matches** in ridgeline patterns between two prints are called **points of similarity**. For a conclusive match between two prints, a minimum of six points of similarity is usually required. Since you are just learning how to do this, **one or two points of similarity per fingerprint** will be sufficient.

Circle each point of similarity you find, mark it with a letter designation, and provide an explanation for each point, as shown in **Figure 9-9**.

Sample print 1

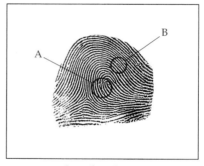
Sample print 2

Explanation
Both prints are right-facing loops

A. Bifurcation forming a broken oval pattern with a complete oval pattern below

B. Oval pattern in ridgeline

**FIGURE 9-9.** How to Label Points of Similarity

**Exercise 9** ■ *Introduction to Forensic Biology* 163

✓ Comprehension Check

Mr. Akumba arrives home and discovers that his back porch window has been forced open. He looks around the house and discovers that his new TV is missing. When the police arrive, they dust the scene for fingerprints. A few days later, Ms. Wright is arrested while breaking and entering another house in the neighborhood. You are the fingerprint examiner for the police department. The chief detective has asked you to compare Ms. Wright's fingerprints with those found at Mr. Akumba's house (**Figures 9-10** and **9-11**).

Left Hand                                                                                          Thumb

Right Hand

Thumb

**FIGURE 9-10.**   Fingerprint File Card for Ms. Wright

**FIGURE 9-11.**   Prints Lifted from Mr. Akumba's Window

1. Do any of the prints from the crime scene match those of the suspect? If so, which prints match which fingers? List the matches below.

2. List the **points of similarity** you used to make your match. Mark each point with a letter on the fingerprint and provide an explanation for each letter (as shown in **Figure 9-9**).

3. **Did Ms. Wright do wrong?** Were those prints at Mr. Akumba's house hers? In a few sentences, summarize the conclusion you reached on the basis of your fingerprint examination.

**Check your answers with your instructor before you continue.**

**Exercise 9** ■ *Introduction to Forensic Biology*

# ACTIVITY 4                                            BLOOD TYPING

Scientists have identified and studied many different human blood groups. Blood groupings are based on the presence of **identifying proteins (also called antigens)** that are incorporated into the cell membranes of all body cells. Although red blood cells have **no personal identifying proteins**, they do have antigens that give us our different blood types. These molecules are referred to as antigens because they are identified as foreign if placed in the body of a person who does not have these molecules.

The most commonly known antigens are those that determine the **A-B-O and Rh** blood groups. These blood groups are well known because they are used to type blood for transfusions and organ transplants. The groups include the following:

- **Type A** blood contains antigen **A** and forms antibodies against **B**.

- **Type B** blood contains antigen **B** and forms antibodies against **A**.

- Type AB blood contains both antigens **A and B**. Type AB **does not form antibodies against either A or B**.

- Type O contains **no A or B antigens** but forms antibodies against **both A and B**.

- The **Rh factor** is the name given to another antigen present in red blood cells. People with **Rh+ blood have the antigen**, while those with **Rh− blood do not**.

Naturally occurring **antibodies (defensive proteins)** in the blood cause a serious transfusion reaction called **agglutination** if a person is transfused with blood containing **foreign** antigens of the A-B-O or Rh blood groups. Babies are not born with these specialized antibodies, but they begin to develop them a few months after birth.

**The combination of an antigen and its corresponding antibody produces agglutination (clumping of blood cells).** The agglutination (clumping) reaction makes it possible to determine blood type.

For example, type A blood has the A antigen. If we add **anti-A** (an antibody that reacts to antigen A) to this blood sample, clumping will occur. The addition of **antibody-B**, however, would not cause these cells to clump. By analyzing the reactions of an unknown blood sample with specific antibodies, the blood type can be determined.

1. Work **in groups**. Get the following supplies: **a spot plate, a wax pencil, some toothpicks, a magnifying glass, and dropper bottles of the following: AB+ and A− blood, anti-A serum, anti-B serum, and anti-Rh serum**.

> **Note:**
> These tests are being carried out with *simulated blood*, **which is nonbiological, and nontoxic.**

| TABLE 9-1 CLUMPING REACTIONS THAT OCCUR IN BLOOD TYPING | | |
|---|---|---|
| BLOOD TYPE | EFFECT OF ADDING ANTIBODY-A | EFFECT OF ADDING ANTIBODY-B |
| A | Clumps form | No clumps |
| B | No clumps | Clumps form |
| AB | Clumps form | Clumps form |
| O | No clumps | No clumps |

2. You will be testing each blood sample for the presence of the antigens A, B, and Rh.

   With the wax pencil, label **three horizontal rows** of your spot plate **A, B, and Rh** as shown in the sample in **Figure 9-12**.

3. You will be practising on two blood samples of **KNOWN type (AB+ and A–)** With these two samples as controls, you will be able to see exactly what the clumping patterns look like for each possibility of A, B, and Rh antigens.

   Label the **first two vertical rows** of your spot plate **AB+** and **A–** as shown in **Figure 9-12**.

4. Place one large **drop of anti-A solution** into each of the **depressions in the row labeled A (one for the known sample of AB+ blood and one for the known sample of A– blood)**.

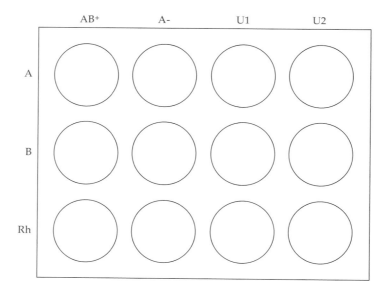

**FIGURE 9-12.** Sample Spot Plate

**Exercise 9** ■ *Introduction to Forensic Biology* 167

Place one large **drop of anti-B solution** into each of the **depressions in the row labeled B.**

Place one large **drop of anti-Rh solution** into each of the **depressions in the row labeled Rh.**

5. For **each of the three depressions for known sample AB+,** put a drop of blood from the **dropper bottle labeled AB+.**

   Stir each sample with a toothpick. **Use a CLEAN TOOTHPICK for each sample.**

6. For **each of the three depressions for known sample A−,** put a drop of blood from the **dropper bottle labeled A−.**

   Stir each sample with a toothpick. **Use a CLEAN TOOTHPICK for each sample.**

7. Observe the mixtures in the six depressions.

   If the agglutination test is positive, you will see a collection of small clumps, or particles, in the depression.

   You may find this easier to see by using a magnifying glass or dissecting microscope.

8. Return to the diagram in **Figure 9-12**. With a colored pencil or a highlighter, **color** each depression where you observed a **clumping** reaction.

9. Repeat the same test procedures using the **two unknown blood samples** supplied by your instructor.

   Label the **next two vertical rows** of your spot plate **U1 and U2** to represent the two unknown samples.

> **Note:**
> **Do not empty the liquid from the depressions containing the positive and the negative controls.**

10. The blood type of unknown **sample U1** is _____.

    The blood type of unknown **sample U2** is _____.

    **Explain your answers.**

## Comprehension Check

Owing to a clerical error, several samples of blood stored at the local blood bank may be incorrectly labeled as to the blood type. The three tests listed below were conducted on each sample. Use your knowledge of antigen–antibody reactions to sort the samples into their correct blood types.

**Test 1**   Unknown sample mixed with "anti-A" serum (contains "type A" antibodies)

**Test 2**   Unknown sample mixed with "anti-B" serum (contains "type B" antibodies)

**Test 3**   Unknown sample mixed with "anti-Rh" serum (contains antibodies against the Rh protein)

1. **Unknown Sample A:** When tested, agglutination (clumping) of red blood cells occurred in Tests 1, 2, and 3.

   **Blood Type:** _____

2. **Unknown Sample B:** When tested, no agglutination occurred in any of the three tests.

   **Blood Type:** _____

3. **Unknown Sample C:** When tested, agglutination occurred in Tests 2 and 3, but not with Test 1.

   **Blood Type:** _____

4. **In your OWN words**, explain the chemical reaction that occurs when anti-B serum is mixed with

   a. a type A blood sample:

Exercise 9 ■ *Introduction to Forensic Biology*

b. a type B blood sample:

c. an Rh-positive blood sample:

> **Note:**
> Blood typing can be useful; however, it is not highly informative. Since there are only four blood groups in the A-B-O system, many people share the same type. Even if the blood type of the evidence matches the suspect, it does not prove that this blood came from the suspect. It could be from anyone that has the same blood type. For this reason, forensic scientists have moved away from conventional blood typing toward the more specific DNA typing technology.

Self Test

Fill in the blanks with the choice that is **most appropriate**. Answers can be used **only once**.

a. ridgeline
b. delta
c. whorl
d. tented arch
e. loop
f. bifurcation
g. divergence
h. fingerprint formula
i. points of similarity
j. sweat glands

1. _____ Complete circle formed by ridgelines between two deltas.

2. _____ Two parallel ridgelines spread apart.

3. _____ Ridgeline divides into two or more branches.

4. _____ Curved ridgelines that enter and exit on the same side of the print.

5. _____ Produce fluids that form latent fingerprints.

6. _____ Set of print classifications for each hand.

7. _____ Exact matches of ridgeline patterns between two fingerprints.

8. _____ Fingerprint pattern in which ridges meet at the center to form a peak.

9. Why do your fingers leave prints when you touch something?

10. Why is it unlikely that identical twins would have identical fingerprints?

Identify the most likely blood type on the basis of the results of the following antigen–antibody reactions. **Explain each answer.**

11. When tested, agglutination occurred when exposed to anti-A serum, but not with anti-B or anti-Rh serums.

12. When tested, agglutination occurred when exposed to anti-A serum, and also with anti-B serum. No agglutination occurred with anti-Rh serum.

13. When tested, agglutination did not occur when exposed to both anti-A and anti-B serums. Agglutination did occur with anti-Rh serum.

EXERCISE 10

# The Circulatory System

## Objectives
After completing this exercise, you should be able to:
- diagram the pattern of blood circulation around the body and label the blood in each location as oxygenated or deoxygenated
- differentiate between the pulmonary and systemic circuits
- identify and explain the function of each of the major structures of the heart
- discuss the connection between blood vessel diameter and the rate of blood flow
- discuss the effect of exercise on pulse rate, blood pressure, and the efficiency of oxygen transport
- discuss the ways in which exercise changes the distribution of blood in the body
- apply your knowledge of vessel and heart structure to cardiovascular health issues.

## CONTENT FOCUS

We hear a lot about exercising our muscles and keeping fit. Activities that increase cardiovascular fitness are an important part of any exercise program. Many people don't realize that the heart is a muscular pump, and so, as with all our body muscles, heart function can be improved by exercise and activity.

The average heart is only about the size of your fist. It normally beats about **72 times per minute.** This may not sound too impressive, but each day, the heart pumps **1500 gallons** of blood. Over your lifetime, this adds up to enough blood to fill **13 supertankers!**

Each beat of the heart **transports blood** that contains **food, oxygen,** and **hormones** to all the cells of the body and **removes wastes** such as **carbon dioxide.**

The circulatory system of humans and most other animals consists of **three basic elements: a pump, blood vessels, and blood (the circulatory fluid).** In this exercise, you will take a closer look at the heart and do some experiments to demonstrate how the circulatory system adjusts to the needs of the body.

## ACTIVITY 1

## CIRCULATION OF BLOOD AROUND THE BODY

Preparation

The heart is divided into **four** chambers. The two **upper** chambers, called **atria,** receive blood returning to the heart. The two **lower** chambers, called **ventricles,** pump blood out.

1. The circulatory pathway that carries blood through the lungs and back to the heart is called the **pulmonary circulation.** The circulatory pathway that carries oxygenated blood through the upper and lower body and back to the heart is called the **systemic circulation.** Following these pathways, you can see that **blood returns to the heart TWICE as it circulates around the body.**

   As shown in **Figure 10-1, blood returns from the body** into the **right atrium.** The blood then travels into the **right ventricle,** which pumps it out to the **lungs.** Blood

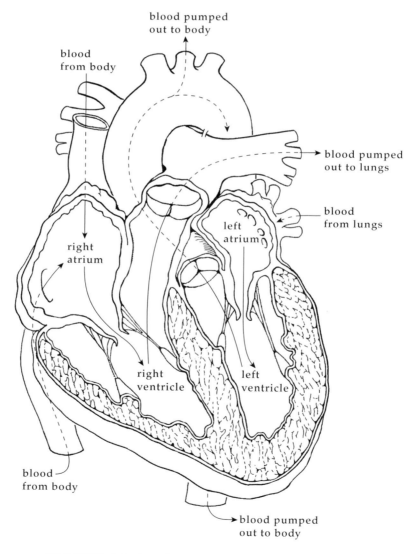

**FIGURE 10-1.** Circulation of Blood Around the Body

Exercise 10 ■ *The Circulatory System* 173

returns from the lungs into the **left atrium,** then goes into the **left ventricle,** which pumps it to the **rest of the body.**

2. Using **a colored pencil or highlighter** (and the information you studied about lung function), color in **blue** each **heart chamber in Figure 10-1** that contains **deoxygenated** blood (blood **low** in oxygen). Color in **red** each **heart chamber** that contains **oxygenated** blood (blood **high** in oxygen).

3. Within the pulmonary and systemic circulatory systems, the blood travels through **arteries, veins,** and **capillaries.**

   **Arteries** always transport blood **away from the heart. Veins** carry blood **toward the heart.** Between the arteries and veins lie **beds of capillaries** where cells pick up oxygen and release carbon dioxide.

   On **Figure 10-2, place a letter "A" next to** each blood vessel that is an **artery.** Place a **letter "V"** next to each **vein.** Place a **letter "C"** next to each **capillary bed.**

4. On **Figure 10-2,** color in **blue** each **heart chamber and blood vessel** that contains **deoxygenated** blood.

   Color each **heart chamber and blood vessel** that contains **oxygenated** blood in **red.**

5. Using the **clues** below, correctly **label** the listed blood vessels on **Figure 10-2.**

| BLOOD VESSEL | CLUE FOR IDENTIFICATION |
|---|---|
| Superior vena cava | Vessel that returns blood from the head to the heart |
| Inferior vena cava | Vessel that returns blood from the lower body to the heart |
| Pulmonary artery | Vessel that carries blood to the lungs |
| Pulmonary vein | Vessel that returns blood from the lungs to the heart |
| Aorta | Vessel that distributes oxygenated blood around the body |
| Carotid artery | Branch off the aorta that delivers blood to the head |

✓ Comprehension Check

1. **(Circle one answer.)** A drop of blood in the **pulmonary artery** is **oxygenated/deoxygenated.**

   Which heart chamber did it come from? _____

   Where will this drop of blood be found next? _____

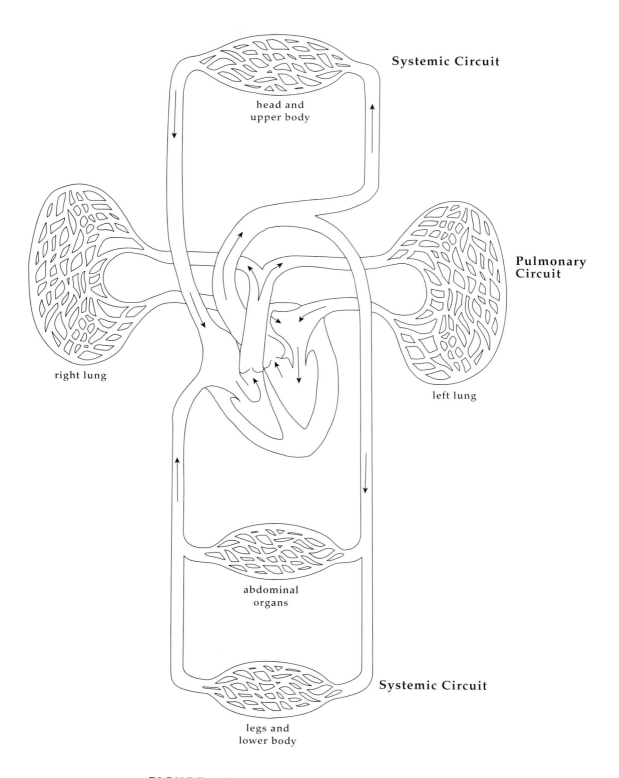

**FIGURE 10-2.** Pulmonary and Systemic Circulation

**Exercise 10** ■ *The Circulatory System*

2. **(Circle one answer.)** The blood flowing through the **left side** of the heart is **oxygenated/deoxygenated**. **Explain your answer.**

3. **(Circle one answer.)** A drop of blood in the **inferior vena cava** is **oxygenated/deoxygenated**.

   Which heart chamber will it enter next? _____

   Where will this drop of blood travel next? _____

4. Do all arteries carry **oxygenated blood? Explain your answer.**

Check your answers and Figure 10-2 labels with your instructor before you continue.

## ACTIVITY 2  A CLOSER LOOK AT THE HEART

1. Work **in groups**. Get a **model of the human heart**. Using **Figure 10-3** as a guide, locate the **four chambers** on the **heart model**. **Observe** the muscular walls of the left and right ventricles on the **heart model**.

   On which side of the heart is the muscular wall the **thickest?** _____
   **Why?**

> **Hint:**
> Review the information in Figures 10-2 and 10-3.

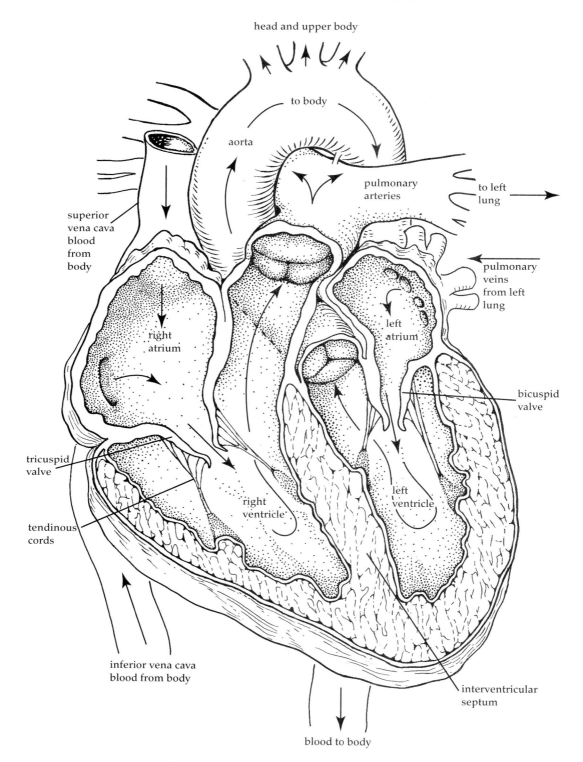

**FIGURE 10-3.** Interior View of the Heart

**Exercise 10 ■** *The Circulatory System*

2. **Heart valves** prevent blood from flowing **backward** through the heart. Each valve section is anchored to the heart wall by **tendinous cords.** These strong cords hold the valve sections **closed** when the ventricles contract.

   **On your model** of the heart, locate **one valve between the right atrium and the right ventricle** (the **tricuspid valve**) and a **second valve** between the **left atrium and the left ventricle** (the **bicuspid** or **mitral valve**).

3. To beat properly, the muscles of the heart must get an adequate blood flow.

   Look at the **outer surface** of the heart model. You will see many **small arteries and veins** branching across the surface of the heart.

   The blood that flows through the heart is moving too fast to be used as a source of oxygen and nutrients by cardiac muscle cells.

   A branch off the **aorta** connects to the **coronary arteries,** which leads to capillary beds in the heart muscle. Blood is returned through the **coronary veins** to the inferior vena cava.

## Comprehension Check

1. **Circle the correct answer** in each of the following statements.

   a. The **bicuspid** valve is **opened/closed** when the **left atrium** contracts.

   b. The **tricuspid** valve is **opened/closed** when the **right ventricle** contracts.

   c. The **tendinous cords** of the **tricuspid** valve are **tightened/relaxed** when the right ventricle contracts.

2. Blood that leaves each ventricle is pumped into an artery. Another valve prevents blood from flowing backward into the ventricle. These are called the **pulmonary and the aortic semilunar valves. Locate** the semilunar valves on the heart model (see **Figure 10-4**).

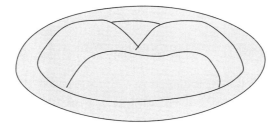

**FIGURE 10-4.** Semilunar Valve

3. The semilunar valve on the **right** side of the heart is between **which heart chamber and which blood vessel?**

   _____      _____

4. The semilunar valve on the **left** side of the heart is between **which heart chamber and which blood vessel?**

   _____      _____

5. (Circle one answer.)

   a. When the right **ventricle** is **contracting,** the **pulmonary** semilunar valve is **opened/closed.**

   b. If a drop of blood is going **toward the lungs,** it has passed through the **pulmonary/aortic** semilunar valve.

6. *Challenge Question!* When the right **atrium** is **contracting,** the **pulmonary** semilunar valve is **opened/closed.**

**Check your answers with your instructor before you continue.**

Exercise 10 ■ *The Circulatory System*

## ACTIVITY 3 — BLOOD VESSEL DIAMETER AND BLOOD FLOW

*Preparation*

The ventricles contract to pump blood through all the arteries of the body. As blood travels away from the heart, it passes through smaller and smaller blood vessels.

How hard does the heart have to work to pump blood through blood vessels of different sizes?

How much force the heart exerts to pump blood **depends on how easily fluid passes through the blood vessel.** Using **rubber tubing to simulate blood vessels,** you will design an experiment to determine whether the **diameter of blood vessels** affects the **rate of blood flow.**

During your experiment, you want to consider **only** the effect of blood vessel diameter. For this reason, you will **not** use a pump to simulate heart action. Blood will flow through your simulated blood vessels using only the force of **gravity.**

1. Work **in groups**.

    Get the following supplies: **one large container with spigot; one stopwatch; one plastic bucket; one large graduated cylinder;** and **three pieces of plastic tubing** (see sizes below):

    | | |
    |---|---|
    | Large | 5/16" inside diameter |
    | Medium | 4/16" inside diameter |
    | Small | 3/16" inside diameter |

2. **Develop a hypothesis** about the relationship between the **blood vessel diameter** and the **rate of blood flow.** Write it in the space below.

 **Check your hypothesis with your instructor before you continue.**

3. Carefully **develop a plan** about how you will get the information you need to test your hypothesis. **List all the steps** of your plan, **including** the **equipment** you will use.

| STEPS OF YOUR EXPERIMENTAL PLAN |
|---|
|  |

4. In the space below, **make a table or chart** that shows your results clearly and neatly. Collect your experimental data and **record the results**.

5. **Plot a graph of your results in Figure 10-5.**

**FIGURE 10-5.** _____

6. Discuss the results with your group members. Write a couple of sentences that **summarize** your **results**.

| Hint: |
|---|
| **Remember—results include only facts, never opinions.** |

7. **Write a conclusion based on your hypothesis and collected data. Support your conclusion** by mentioning facts collected during your experiment.




8. **(Circle one answer.)** As blood travels from arteries into arterioles, the rate of blood flow will **increase/decrease**.

9. Capillaries are the **smallest** blood vessels. How does this fact affect the **rate** of blood flow?




10. Why is **flow rate** an important factor for determining how **efficiently** cells can pick up oxygen and remove carbon dioxide?




11. In the disease **atherosclerosis**, fatty plaques accumulate on the inside of artery walls. What effect would this have on the **rate** of blood flow? **Explain your answer.**

**Check your answers with your instructor before you continue.**

**Exercise 10** ■ *The Circulatory System* 183

# ACTIVITY 4                                         THE EFFECT OF EXERCISE

*Preparation*

Every time the ventricles contract **(systole)**, forcing blood out of the heart, pressure in the blood vessels **increases**. When the ventricles **relax (diastole)**, pressure in the blood vessels **drops** again. The pressure exerted by the blood on the walls of the blood vessels is called **blood pressure**. The original instrument used to measure blood pressure forced a **column of mercury** upward in a glass tube. Newer instruments that measure blood pressure still use the same measurements **(millimeters of mercury or mm Hg)**. The average blood pressure in a young adult male is approximately **120 mm Hg during systole, and 80 mm Hg during diastole**. This is expressed as **120/80**. Young adult females average about 8–10 mm Hg less than males. Blood pressure can be lower in adults that exercise regularly.

Blood is forced into your arteries during each contraction of the heart. You can feel this wave of blood moving through the arteries as a **pulse** in the carotid artery of your neck. The pulse can also be felt in the radial artery of your wrist and other locations in the body. The **number of pulses** tells you **how fast the heart is beating**. The average pulse rate in a young adult is **72 beats per minute**.

1. Working with a **group**, do the following experiments.

   One student will be the test subject, another will monitor the blood pressure, another will monitor the pulse rate, and the fourth will be the timekeeper and record the experimental data.

> ### *Caution!*
> **DO NOT be the test subject for this activity if you have heart or blood pressure problems!**

2. Get the following supplies: **one blood pressure monitor, a stopwatch, and an aerobic step**.

> ### *Caution!*
> **Blood pressure monitors are fragile and expensive! Please be careful with this equipment.**

3. Sit quietly for **one minute**. While sitting, take your **resting pulse rate and blood pressure**.

   Resting pulse rate: _____

   Resting blood pressure: _____

4. **Rapidly step up and down** on the aerobic step for **five minutes.** You should really be working out!

   **Immediately** sit down and record the **pulse rate and blood pressure.**

   Exercise pulse rate: _____

   Exercise blood pressure: _____

5. Is there a difference in pulse rate before and after exercise? _____

   If so, **describe the difference.**

6. Is there a difference in blood pressure before and after exercise? _____

   If so, **describe the difference.**

7. What is the heart doing when **blood pressure** increases? _____

   What is the heart doing when the **pulse rate** increases? _____

8. How do these **two different changes in heart action** help the body during exercise? In your answer, use the following terms: **blood flow, oxygen, carbon dioxide, cell respiration,** and **ATP.**

**Exercise 10** ■ *The Circulatory System*

Self Test

1. Follow a drop of blood around the body. Begin with the tissue capillaries supplying your left big toe. **Place these locations in the correct sequence.**

    a. _1_  Tissue capillaries in big toe
    b. ___  Pulmonary artery
    c. ___  Left atrium
    d. ___  Pulmonary vein
    e. ___  Right atrium
    f. ___  Left ventricle
    g. ___  Inferior vena cava
    h. ___  Aorta
    i. ___  Right ventricle
    j. ___  Arterioles (small arteries)
    k. ___  Venules (small veins)
    l. ___  Lungs

2. If there were **no tendinous cords** attached to the tricuspid valve, and the right ventricle is contracted, where would the blood go?

3. If a person's pulse rate is 96 pulses per minute, what is this person's **heart rate**?

4. For the following body locations, enter **"D"** if the blood is **deoxygenated** and **"O"** if the blood is **oxygenated.**

   ___ Tissue capillaries entering big toe

   ___ Pulmonary artery

   ___ Left atrium

   ___ Pulmonary vein

   ___ Right atrium

   ___ Left ventricle

   ___ Inferior vena cava

   ___ Aorta

   ___ Right ventricle

   ___ Tissue capillaries leaving big toe

   ___ Arterioles (small arteries)

   ___ Venules (small veins)

5. You are a doctor listening to heart sounds through a stethoscope. While listening, you notice an unusual "hissing" sound in a patient's heart. On consulting the patient's medical records, you find that as a child he suffered from rheumatic fever, which causes scar tissue to form around the heart valves. What might be causing the hissing sound?

6. When a person **blushes,** blood flow to the skin increases. In order for this change in blood flow pattern to occur, how must the **diameter** of skin arteries change?

Exercise 10 ■ *The Circulatory System*

7. Place an "X" in front of the **one** measurement of oxygen concentration **that is most likely to be correct for all four locations.**

| X | RIGHT VENTRICLE | PULMONARY ARTERY | PULMONARY VEIN | LEFT ATRIUM |
|---|---|---|---|---|
|   | 40 | 40 | 100 | 100 |
|   | 100 | 40 | 40 | 100 |
|   | 100 | 100 | 100 | 100 |
|   | 40 | 40 | 40 | 40 |
|   | 40 | 100 | 40 | 100 |

The graph in **Figure 10-6** shows the distribution of blood (by percentage) in the circulatory system when the body is at rest and during heavy exercise.

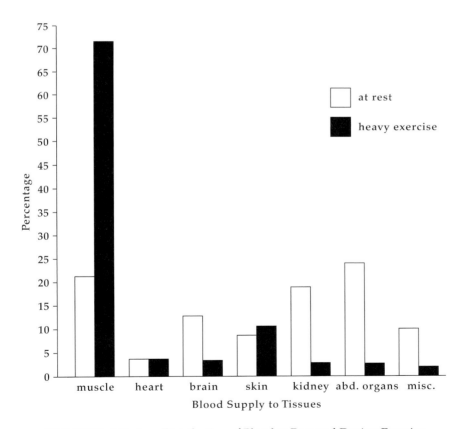

**FIGURE 10-6.** Distribution of Blood at Rest and During Exercise

8. In reference to the graph in **Figure 10-6**, what **change** occurs to the **percentage of blood supply to the muscles** when you are exercising?

   Considering the **change in the percentage of blood flow** to the muscles shown in the graph, what probably happens to the **diameter of the blood vessels** that supply blood to the muscle tissue during exercise?

9. What **change** occurs in the percentage of blood supply to the **abdominal organs** during exercise?

   Considering the **change in the percentage of blood flow** to the abdominal organs shown in the graph, what probably happens to the **diameter of the blood vessels** that supply blood to the abdominal organs during exercise?

10. What **change** occurs in the **percentage of blood supply to the skin**?

    How is this **change** in blood flow **helpful to the body during exercise**?

EXERCISE 11

# Introduction to Anatomy: Dissecting the Fetal Pig

## Objectives
After completing this exercise, you should be able to:
- identify and compare the external anatomical features of the male and female fetal pig
- identify and explain the function of each of the major structures of the mouth, the neck, and the thoracic cavity
- explain how organ structures are specialized to perform specific functions and give examples
- compare and contrast pig anatomical features with human anatomical features
- discuss the difference between lung capacity with normal and deep breathing.

## CONTENT FOCUS

During the next several weeks, you will be studying mammalian anatomy and physiology. To help you get a clear idea of the various organ systems and how they work, you will be looking at the anatomy of the **fetal pig**.

**Why fetal pigs?** It may surprise you to learn that humans and pigs are very similar in anatomy. Today, pig organs and skin are frequently used for transplants and grafts to replace damaged human tissues. In addition, the fetal pig skeleton is not fully calcified, making dissection easier to perform. During extreme weather, farmers often sell surplus animals rather than run the risk of them dying from heat or cold. As part of the butchering process, all the organs, including the uterus, are removed. If fetal pigs are found in the uterus, they are preserved for educational purposes.

## ACTIVITY 1 — GETTING STARTED

1. Get the following supplies: **two long pieces of string, a tray, several paper towels, and some dissecting instruments (one scalpel, one pair of scissors, one large pair of forceps, and one blunt probe).**

2. Your laboratory instructor will distribute fetal pigs to each group or give you instructions for obtaining a pig.

3. Spread **two paper towels** on the tray. **Place the pig on its back on the tray.**

4. Tie the **ends** of the first string **tightly** around the two front hooves of your pig and slide it **underneath the tray.** Tie the **ends** of the other string around the two rear hooves and slide it underneath the tray.

5. Rotate the tray so that the pig's **tail is facing you.**

   Where is the **left side of the pig?** _____

> **Note:**
> Keep the position of the pig in mind as you proceed with the dissection. The instructions refer to the **pig's right and left sides, NOT your right and left sides.**

## ACTIVITY 2 — FOLLOWING ANATOMICAL DIRECTIONS

*Preparation*

Before we begin the dissection, it is necessary to understand some special terms that relate to anatomical locations and directions in the pig's body. Using **Figure 11-1** as a guide, use the correct anatomical terms to describe the location of the following:

1. The pig, as placed on the tray, is lying on its _____ surface.

2. The umbilical cord is located on the _____ side of the pig.

3. The attachment of the umbilical cord to the belly is _____ to the **hind** legs.

4. The attachment of the umbilical cord to the belly is _____ to the **front** legs.

Exercise 11 ■ *Introduction to Anatomy: Dissecting the Fetal Pig*   191

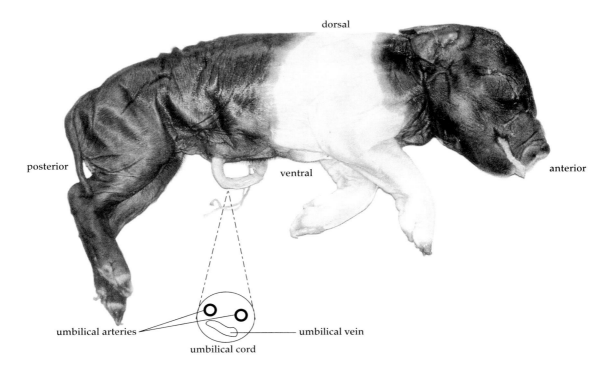

**FIGURE 11-1.** Anatomical Locations in the Pig

## ACTIVITY 3      EXTERNAL ANATOMY OF THE FETAL PIG

1. Determine the **sex** of your pig, using **Figures 11-2 and 11-3** as a guide. There are two ways to tell males from females **externally**.

> **Note:**
> Make sure you observe pigs of both sexes.

2. Look for the **anus**, which is located close to the base of the tail.

    **Female pigs** have a small, fingerlike projection, **just ventral to the anus**. This projection is called the **genital papilla**. On either side of the papilla are the two **labia**. The **urogenital opening** is located between the two labia. The term **urogenital** refers to the double function of this opening. It allows for the excretion of urine ("uro" refers to the urinary system) and also leads to the **reproductive structures** ("genital" refers to the reproductive system).

> **Hint:**
> If the papilla is not evident, you have a male pig.

3. In **male pigs,** you will notice the skin appears puffy or baggy just **anterior to the anus.** This is the **scrotum**, which contains the **testes**.

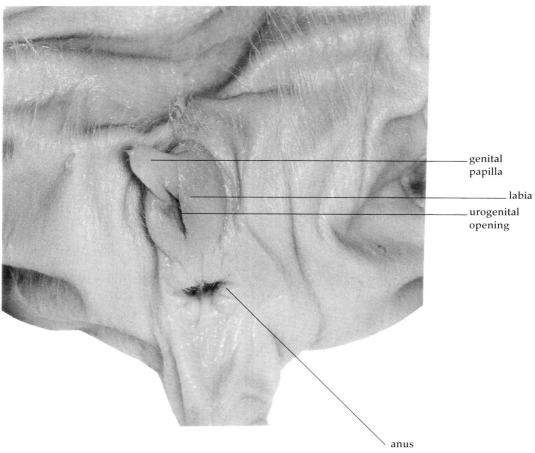

**FIGURE 11-2.** Female External Anatomy

**Exercise 11** ■ *Introduction to Anatomy: Dissecting the Fetal Pig* 193

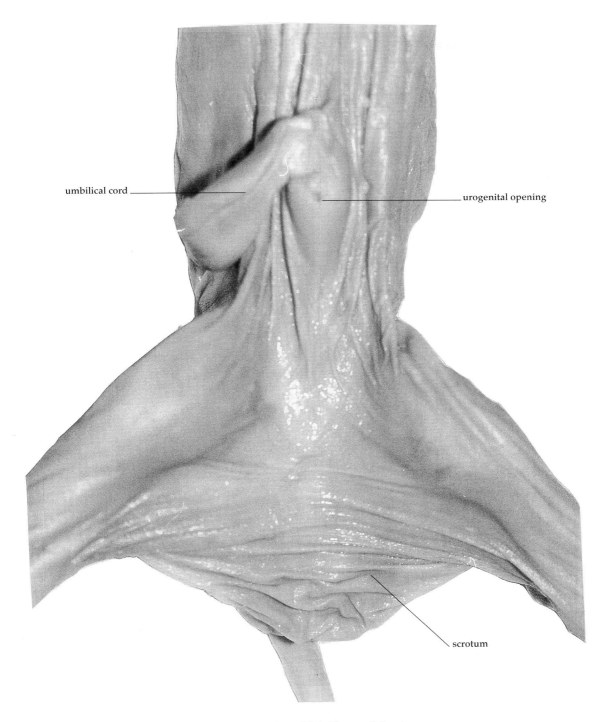

**FIGURE 11-3.** Male External Anatomy

4. The **urogenital opening** in the male is located just **posterior** to the attachment of the **umbilical cord** to the body wall. As with the female pigs, this opening functions in both the urinary and reproductive systems. Both urine and sperm are released here, but not at the same time.

   You cannot see the **penis** on the exterior of the body because pigs, like many other four-legged animals, have a **retractable** penis, which can be seen when the animals reproduce. You will notice that both male and female pigs have two rows of **nipples** on either side of the umbilical cord. As with humans, the nipples are only functional in females.

5. The body of the pig is divided into three regions: **head, neck,** and **trunk.** The trunk region, with the four legs and tail attached, is further subdivided into the **anterior** area, the **thorax** (chest area), and the **posterior** area, the **abdomen.** The thorax contains the heart and the lungs, enclosed by a protective rib cage. The abdomen contains the organs of the digestive system, along with organs of many other important systems.

   The **umbilical cord** attaches the fetus to the **placenta** in the mother.

6. Each pig has a **slit** cut in the side of the neck. This cut was made to inject the circulatory system with **colored latex** (a form of liquid rubber). The **arteries** have been injected with **pink latex** and the **veins** have been injected with **blue latex.** This will help you tell the blood vessels apart when you study the circulatory system.

# ACTIVITY 4                       DISSECTION OF THE MOUTH

1. Gradually, cut through **both sides of the jaw,** as illustrated in **Figure 11-4.** As you cut, **alternate** between the right and left sides until you can open the mouth wide.

> **Caution!**
> **Young pigs have very sharp, pointed teeth. Be careful where you put your fingers while opening the mouth.**

2. Look at the **roof** of the mouth and **feel its texture.** The rippled **anterior region** is hard because it is composed of **bone and cartilage.**

   This area is called the **hard palate.** Notice that the hard palate is divided into **two halves.** The line separating the left and right sections of the hard palate runs directly along the **body midline.**

**Exercise 11** ■ *Introduction to Anatomy: Dissecting the Fetal Pig*     **195**

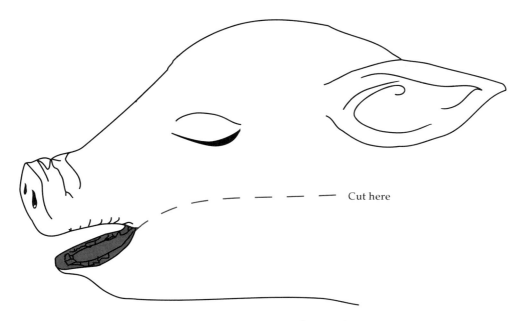

**FIGURE 11-4.** Pattern for Jaw Incisions

**Using** a pencil, label the **hard palate** and the **midline** on the diagram in **Figure 11-5**.

3. The **posterior region** of the roof of the mouth is the **soft palate.** Feel the texture of this area. **How is it different** from the hard palate area?

4. The hard and the soft palate separate the mouth from the **nasal cavity.** As you may have experienced, drainage from the nose can enter the mouth, so there must be a **common area** where these two regions come together. This is the **pharynx.**

   The pharynx can be seen **posterior** to the soft palate.

   Insert the **tip of your blunt probe** into the opening of the pharynx and probe **anteriorly underneath the soft palate. Your probe is now in the nasal cavity.**

   Using a pencil, **add labels** for the **pharynx** and **soft palate** to Figure 11-5.

5. Pull the mouth completely open and look down into the throat. There is a hood-shaped flap of tissue called the **epiglottis.** The epiglottis prevents food and liquid from entering the **glottis,** the opening to the air passageways (the prefix **"epi"** means *outside*).

   **Using** a pencil, **add labels** for the **epiglottis** and **glottis** to Figure 11-5.

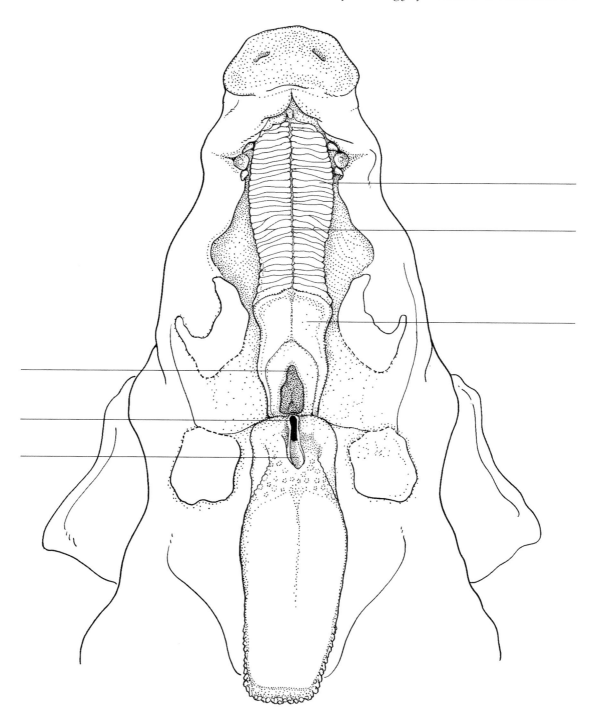

**FIGURE 11-5.** Structures of the Mouth

Check your Figure 11-5 labels with your instructor before you continue.

**Exercise 11** ■ *Introduction to Anatomy: Dissecting the Fetal Pig* 197

6. If you are eating and your food "goes down the wrong pipe," **what pipe** did the food enter? _____

## ACTIVITY 5                                      DISSECTION OF THE NECK

> ### Caution!
> While cutting, be careful to keep the point of the scissors away from underlying organs.

1. **Using the dotted lines on Figure 11-6** as a guide, use **scissors** to cut through the skin and muscle layers. Begin at the **chin whiskers (number 1)** and continue to the **anterior part of the sternum or breastbone (number 2).**

2. Make two additional incisions **across the throat (number 1 to number 3).**

3. Use your **blunt probe** to separate the muscles from the organs in the neck region. You will expose the **thymus gland** on both sides of the neck. The thymus gland has an important role in **immune system function.**

4. With your fingers, feel around at the **anterior end** of the dissected area **(toward number 1 on Figure 11-6),** until you locate a hard bulblike structure. This is the **larynx.**

   **Remove the muscle tissue** that is covering the larynx until it can be clearly seen. The larynx is sometimes known as the **"voice box."**

   Considering the **function** of the larynx, what vibrates inside this structure? _____

5. Continue to expose the structures connected to the **posterior end of the larynx.** Remove the overlying tissues until you can see the tubular **trachea,** commonly known as the **windpipe.**

6. The trachea is supported along its length by many **cartilage rings.** The trachea's structure, with the cartilage rings, is similar in design to a vacuum cleaner hose.

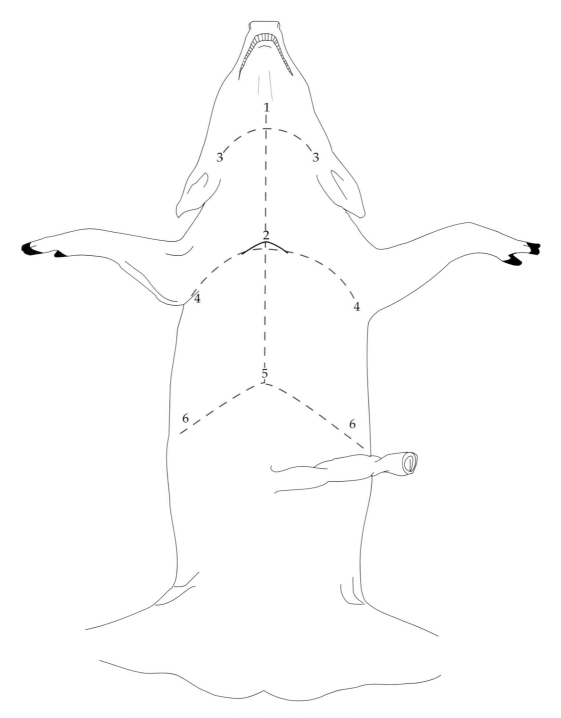

**FIGURE 11-6.** Pattern for Body Cavity Incisions

**Exercise 11** ■ *Introduction to Anatomy: Dissecting the Fetal Pig*

What **function** do the metal rings serve in the vacuum cleaner hose?



What is the **function of the trachea** in your body?



Why is the **presence of cartilage rings** in the trachea important for your survival?



7. At the **posterior end of the trachea,** you will see a small, dark-brown organ, the **thyroid gland.** The thyroid gland produces **thyroid hormone,** which controls the body's **metabolic rate.**

8. Using your **blunt probe,** carefully separate the connective tissue from the sides of the trachea. The esophagus is **dorsal to the trachea.**

   Feel underneath the trachea and **lift the esophagus** with your probe. How is the esophagus **different in structure** from the trachea?



9. Is the esophagus part of the respiratory system? _____ **Explain your answer.**



■ **Check your answers with your instructor before you continue.**

10. On the side of the neck **opposite to the slit,** carefully remove the overlying muscle and connective tissue to expose the **carotid artery** and **two jugular veins.**

    These vessels will be running **parallel** to the trachea (see **Figure 11-7**).

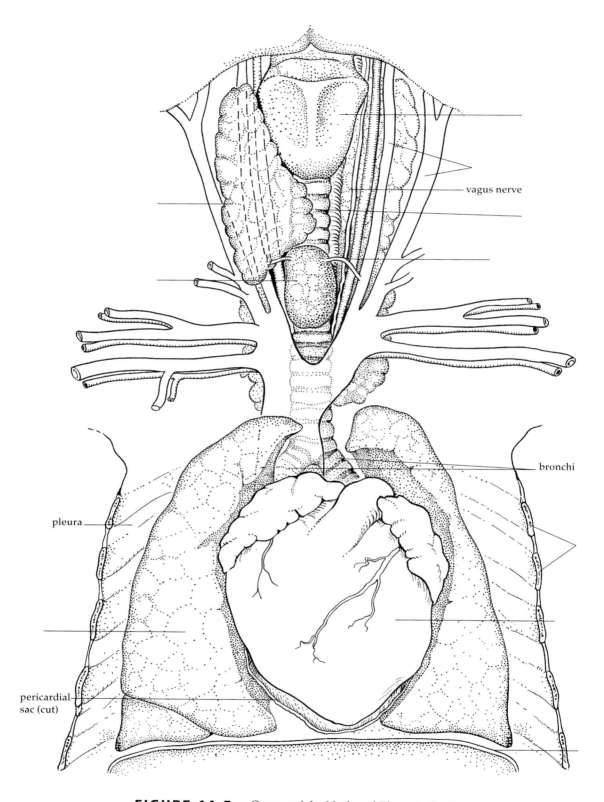

**FIGURE 11-7.** Organs of the Neck and Thoracic Cavity

**Exercise 11** ■ *Introduction to Anatomy: Dissecting the Fetal Pig*

11. The **carotid artery** is the **closest blood vessel to the trachea** and should be filled with **pink** latex. The **two jugular** veins should appear **blue**. The carotid artery carries blood rich in oxygen and nutrients toward the head. The jugular veins carry deoxygenated blood back toward the heart.

12. *Dissection Challenge!* Can you find the **vagus nerve** in the neck? It looks like a **flat,** white thread running close to the carotid artery and internal jugular vein. The vagus nerve is crucial for maintaining the normal state of homeostasis. Among other functions, this nerve **regulates heart rate, breathing, and digestive system activity.**

13. **Label** the following structures on **Figure 11-7: thymus, larynx, trachea, thyroid gland, carotid artery, and jugular veins.**

### Comprehension Check

Fill in the blanks with the choice that is **most appropriate** for the listed functions. Answers can be used **only once.**

a. Carotid artery
b. Esophagus
c. Larynx
d. Trachea
e. Thyroid gland
f. Vagus nerve
g. Thymus
h. Jugular vein

1. _____ Tube that transports food to the stomach.

2. _____ A blood vessel carrying deoxygenated blood back to the heart.

3. _____ This structure produces a hormone that controls the body's metabolic rate.

4. _____ Tube held open by cartilage rings.

5. _____ A blood vessel carrying nutrient-rich blood to the head.

6. _____ This structure contains the vocal cords.

7. _____ Plays an important role in immune system defense.

8. _____ Regulates heart rate, breathing, and digestive system activities.

## ACTIVITY 6 — DISSECTION OF THE THORACIC CAVITY

1. With your fingers, find the **anterior end of the sternum**. With your **scissors**, cut through the **sternum and the rib cage** until you reach the end of the rib cage (cut from **number 2 to number 5** and from **number 2 to number 4** on **Figure 11-6**). This will give you a clearer view of the ribs and help you pinpoint the site of your next incision.

2. **Make two more incisions** toward the sides of the body to completely open up the chest cavity **(number 5 to number 6)**. As you cut, you will be separating the **ribs** from the **diaphragm**, a large muscle that extends **completely across** the body cavity. **The diaphragm expands the chest cavity so that air flows into the lungs.**

    The thoracic cavity is lined with a smooth membrane.

3. As you look into the thoracic cavity, the **two lungs** are easy to see. Each lung is enclosed in a **pleural cavity**, lined by the **pleural membranes**.

    The lungs are composed of millions of tiny sacs which are the sites of gas exchange for the whole body. **Oxygen enters the lungs and carbon dioxide is removed.**

> **Note:**
> The trachea extends posteriorly from the larynx and divides into branches called bronchi, which extend to the lungs. The branches cannot be seen at this stage of the dissection, but we will examine them later.

4. Located between the two lungs is the **heart**. The heart is a muscular organ that **pumps blood to the lungs and the body tissues**. The pig heart is located in the center of the chest cavity. This is also true of the human heart (not on the left side as is commonly believed).

    The **pericardium** (also called the **pericardial sac**) is a tough membranous sheet that completely encloses the heart. **Remove the pericardial sac** so that you can get a better look at the heart.

5. **Add the following labels** to the diagram in **Figure 11-7**: **lungs, heart, ribs**, and **diaphragm**.

**Check your Figure 11-7 labels with your instructor before you continue.**

**Exercise 11** ■ *Introduction to Anatomy: Dissecting the Fetal Pig*

6. At the completion of your dissection, do the following:

   a. Get a plastic storage bag for your pig. Place a strip of tape near the bottom of the bag and write your group member's names on the tape.

   b. Slide the pig off the tray with the string still attached to the limbs.

   c. Wrap the pig in damp paper towels and seal it in the storage bag. Your instructor will tell you where to store your pig.

   d. **Wash and dry your trays and tools. Your instructor will give you specific instructions for disposing of any remaining pig tissues.**

## ACTIVITY 7 — MEASURING LUNG CAPACITY

Preparation

Just looking at the lungs does not give you a very good idea of how the lungs function. The following activities will allow you to experiment on yourself and your lab partners and develop some ideas about how respiratory function adjusts to your body's needs when you exercise.

The amount of air that enters and leaves your lungs each time you take a breath is called the **tidal volume.** In most men and women, the tidal volume is about **500 cc (half a liter).**

You can increase the amount of air inhaled and exhaled by deep breathing. The **maximum** amount of air that you can move in and out during a single breath is called the **vital capacity.**

1. Work **with a partner. Each** student will perform this experiment and record their **individual** results.

---

### *Caution!*
**DO NOT use the balloon method if you have respiratory or heart problems!**

2. Get a metric ruler, a clamp, an index card, and a balloon for each person.

> **Note:**
> Read through the directions COMPLETELY before beginning!

3. Measure your respiratory volume according to the following directions.

   a. Sit down and relax.

   b. **Inhale normally** and then exhale only that normal breath into the balloon.

> **Note:**
> If you have trouble inflating the balloon, stretch it several times and try again.

   c. **Immediately** twist the balloon several times and **clamp it shut** so that no air escapes.

   d. Place the balloon on its **side** on the laboratory counter. Use the ruler to measure the height of the balloon **in centimeters.**

   Place the ruler near the **center** of the balloon. Don't push down on the balloon when you're measuring.

   An easy way to do this is to place the ruler standing up behind the center of the balloon. Place an **index card** lightly on top of the balloon. Where the **index card meets the ruler,** take your height measurement.

   e. **Record** your results and **repeat** the experiment **twice more. Record** the results **of all three trials in Table 11-1.**

4. Calculate the **average balloon diameter** for your three trials of exhaled air **at rest. Record** the average in **Table 11-1**.

5. a. Take **one very deep breath** and exhale as much air as possible into the balloon. **Clamp** and **measure** the balloon as in step 3 above.

# Exercise 11 ■ *Introduction to Anatomy: Dissecting the Fetal Pig*

| TABLE 11-1 MEASURING LUNG CAPACITY ||
|---|---|
| **DIAMETER (cm) OF BALLOON FILLED BY NORMAL BREATH** ||
| Trial 1 | |
| Trial 2 | |
| Trial 3 | |
| Average | |
| Volume of air (cc) in balloon filled with normal breath: ||
| **DIAMETER (cm) OF BALLOON FILLED BY DEEP BREATH** ||
| Trial 1 | |
| Trial 2 | |
| Trial 3 | |
| Average | |
| Volume of air (cc) in balloon filled with deep breath: ||

    b. Record your results in **Table 11-1**.

6. Calculate the **average balloon diameter** for your three trials of exhaled air **while deep breathing. Record** the average in **Table 11-1**.

7. **Use the graph** in **Figure 11-8** to convert the average diameter of the balloon filled by **a normal breath** and the balloon filled by **a deep breath** into **cubic centimeters (cc) of lung volume.**

   **Record** the information in **Table 11-1**.

8. What is your **tidal volume?** _____

   What is your **vital capacity?** _____

> **Note:**
> Your measurement of tidal volume may be slightly higher than your actual tidal volume, since you also exhaled the reserve air held in your lungs.

9. Which volume of air is exchanged while you are sleeping? _____

   Which volume of air is exchanged while you are running? _____

**FIGURE 11-8.** Relationship Between Balloon Diameter and Volume of Exhaled Air

 **Check your answers with your instructor before you continue.**

**Exercise 11** ■ *Introduction to Anatomy: Dissecting the Fetal Pig*

Self Test

Match the definitions in the right column with the terms in the left column. Each answer can be used only once.

1. _____ Anterior        a. Back

2. _____ Dorsal          b. Tail end of the body

3. _____ Posterior       c. Head end of the body

4. _____ Ventral         d. Belly side

5. Your lab partner arrived late to class on the day you started your fetal pig dissection. You have already selected a male pig for your dissection and you want your partner to choose a female. **Describe** how your partner can recognize and select a female pig from the container. **Be specific.**

6. Erin is playing shortstop on her college baseball team when she is hit in the throat by a line drive. On the way to the hospital, she has a very difficult time breathing. Which of the following is the most likely cause of her problem? **Explain** your answer.

   a. Bruised diaphragm
   b. Concussion
   c. Broken rib cage
   d. Crushed trachea
   e. Blocked esophagus

7. Name **one part of the digestive system that** is located in the **thoracic** cavity.




8. Name a structure that lies **between** the thoracic and abdominal cavities.




9. When breathing, you are never able to completely fill your lungs with fresh air. A certain volume of stale air (with the oxygen removed) always remains in the air passageways. In some lung diseases, such as **emphysema,** large amounts of stale air accumulate and cannot be expelled. The lungs are filled with air that is useless for gas exchange. **Vital capacity** is significantly less than in normal lungs.

   Predict what will happen when a person with emphysema exercises heavily. **Explain** your answer.




10. In regard to your answer to question 9, explain how lack of air affects the body tissues. In your answer, use the following terms: **energy, oxygen, mitochondria, ATP, and red blood cells.**

EXERCISE **12**

# Organ Systems in the Abdominal Cavity

## Objectives
After completing this exercise, you should be able to:
- identify and explain the function of each of the major structures of the abdominal cavity
- explain how the stomach and small intestine are specialized to perform specific functions
- compare and contrast pig anatomical features with human anatomical features
- apply your knowledge of the benefits of increasing surface area in the radish root to similar modifications in the intestine and lung
- relate your observations of earthworm locomotion to peristaltic action in organs of the digestive tract.

## ACTIVITY 1   DISSECTION OF THE ABDOMINAL CAVITY

1. **Get your pig** from the storage container and set it up on a tray as you did in the last laboratory period.

   Get a set of dissecting instruments.

2. Use your **scissors** to cut through the **skin and muscles** in the abdominal region.

   **Pull up on the umbilical cord** to hold the skin away from the abdominal organs and cut from **number 5 to number 7 (see Figure 12-1).**

   Make two more incisions **around the umbilical cord (number 7 to number 8).**

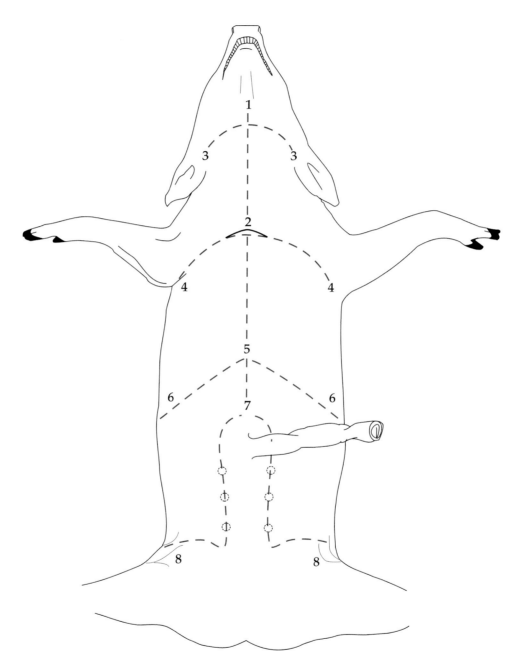

**FIGURE 12-1.** Pattern for Body Cavity Incisions

Exercise 12 ■ *Organ Systems in the Abdominal Cavity*

3. The flap of tissue created by these cuts contains the **umbilical vein, the two umbilical arteries,** and the **urinary bladder.**

   Cut the umbilical vein **close to the umbilical cord.** This will allow you to turn back this flap so that it lies between the hind legs.

> **Hint:**
> If needed, rinse out the entire body cavity and blot it dry with paper towels. This will make it easier to locate and identify the abdominal organs.

4. Just as in the thoracic cavity, the abdominal cavity is sealed with a smooth membrane called the **peritoneum.**

5. The largest organ in the abdominal cavity is the **liver.** It is reddish-brown with several lobes and is located just **posterior to the diaphragm.**

   The liver has many functions, including **storage of energy reserves, detoxification of poisons, and bile formation.**

6. And speaking of bile, lift the **right lobe** of the liver and locate the **gall bladder,** a small, saclike organ.

   The gall bladder **stores bile produced by the liver. Bile aids in fat digestion.** The bile is released through the **bile duct** into the small intestine, where fat digestion takes place.

7. Raise the liver to locate the **stomach,** a large, hollow bag located on the left side of the body.

   **The stomach stores food and releases it gradually into the small intestine.** Digestion of proteins begins here.

8. The long, dark red organ lying to the left of the stomach is the **spleen.**

   The spleen **functions as part of the immune system** and also **removes damaged and worn-out blood cells from circulation.**

   Within the abdominal cavity, the organs are held in position by membranes called **mesenteries.** You can see this membrane attaching the spleen to the stomach.

9. The **pancreas** is located between the stomach and the small intestine. To expose this organ, lift the stomach and use your **blunt probe** to dissect through the membranes in this area. The pancreas is an **elongated, pale, granular** organ positioned **across the body cavity.**

   The pancreas **secretes digestive enzymes** through a duct into the small intestine. It also functions as an **endocrine organ,** producing **two hormones that control blood sugar level.**

10. The **small intestine,** on the right side of the abdominal cavity, is a long, thin tube. Most chemical digestion occurs here. In the small intestine, **enzymes break down proteins, fats, carbohydrates, and nucleic acids.**

    Hold up one of the loops of the small intestine and notice that it is held together by **mesenteries.** If you look closely, you can see **arteries and veins** fanning out across the membrane.

    Mesenteries provide a **support structure** for blood vessels and nerves leading to all the abdominal organs.

11. The **large intestine** is on the left side of the abdominal cavity. The main function of the large intestine is the **reabsorption of water from wastes,** preventing dehydration. The concentrated wastes are stored in the **rectum** and eliminated through the **anus.**

12. Lift the intestines on either side of the body and you will see a **kidney.** The kidneys **remove nitrogen-containing wastes and form urine.**

13. The **urinary bladder,** located between the two umbilical arteries, is a **temporary storage organ for urine.** When the bladder fills, nerve endings in its muscular walls are stimulated, causing the muscles to contract for urination.

14. **Label** the following structures on **Figure 12-2**: **liver, gall bladder, stomach, spleen, pancreas, small intestine, large intestine, mesenteries, kidney, and rectum.**

**Check your figure labels with your instructor before you continue.**

**Exercise 12** ■ *Organ Systems in the Abdominal Cavity*

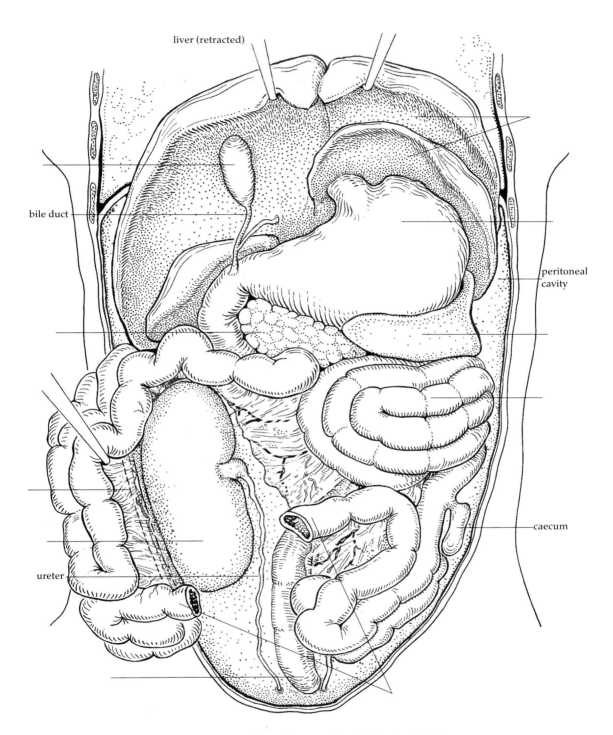

**FIGURE 12-2.** Organs of the Abdominal Cavity

## ACTIVITY 2 — THE URINARY SYSTEM

Preparation

Waste products are removed from the body by several systems. For example, **carbon dioxide,** a waste product of **cell respiration,** is removed by the **lungs. The nitrogen-containing** waste products of **protein digestion** are excreted by the urinary system.

The urinary system includes the two **kidneys,** which **produce urine;** the two **ureters,** which **carry the urine to the bladder;** the **bladder,** which **stores the urine;** and the **urethra,** which **transports urine** out of the body through the **urogenital or urinary opening** (depending on your sex).

The kidneys also have an important function in maintaining **homeostasis** (keeping internal body conditions stable). The body must maintain adequate levels of salts, sugars, and other substances in the bloodstream, while disposing of the excess. The kidneys play a key role in this process, keeping **pH** values, **fluid and salt content,** and **plasma levels of valuable materials** nearly constant.

1. Push the intestinal mass to the **left side** of the abdominal cavity, exposing the **right kidney.** The kidneys are located along the **dorsal wall of the abdominal cavity** (see **Figure 12-3**).

2. You will notice that the kidney is covered by a thick layer of connective tissue, the **peritoneum.**

   **Using the blunt probe,** carefully remove the peritoneum, uncovering the right kidney and ureter.

   In addition to the ureter, you will see the **renal artery and vein.** The renal **artery** carries blood **to the kidney for filtration.** The blood **returns** to the body circulation through the **renal vein.**

3. Each kidney is drained by a **ureter,** a tube leading from the kidney to the urinary bladder.

   Locate the ureter and follow it to the **bladder,** which is located in the flap of tissue created by your **incisions around the umbilical cord in Activity 1.** The bladder can be found **between the two umbilical arteries.**

   The **posterior end of the bladder** narrows to form a tube, the **urethra,** which transports urine out of the body. Most of the urethra cannot be seen, because it is positioned deeper in the pelvic cavity.

4. Label the following structures on **Figure 12-3**: **kidney, urethra, urogenital opening,** and **bladder.**

# Exercise 12 ■ Organ Systems in the Abdominal Cavity

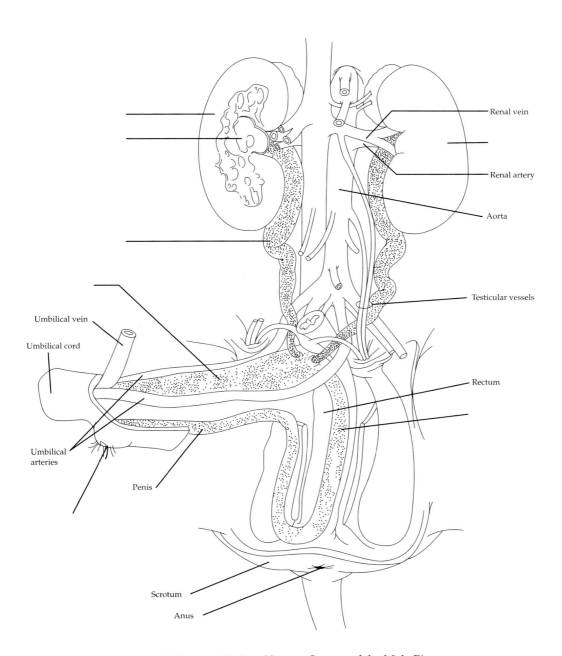

**FIGURE 12-3.** Urinary System of the Male Pig

5. Cut the ureter and renal blood vessels. Remove the intact right kidney from the abdominal cavity. If necessary, remove additional peritoneum so that the kidney can be removed without damage.

6. Make an incision through the long axis of the kidney, cutting it into two equal halves (like opening a book).

   Within the kidney, you will see an **outer area** of densely packed tissue, the **cortex**. Urine formation begins in microscopic structures within this region.

   The **inner** section, the **medulla,** appears fibrous. This area collects the urine and funnels it to the ureter.

   Observe the connection of the ureter to the kidney near the center of this region.

7. Label the following structures on the dissected kidney in **Figure 12-3**: **cortex, medulla,** and **ureter.**

## Comprehension Check

Fill in the blanks with the choice that is most appropriate to describe the function of each part of the urinary system. **Answers can be used more than once. Some questions may have more than one correct answer**.

a. Ureter  
b. Kidney  
c. Urethra  
d. Renal artery  
e. Renal vein  
f. Peritoneum  
g. Urinary bladder  
h. Medulla  
i. Cortex

1. _____ This membrane holds the kidneys in position.
2. _____ If a drop of urine is in the ureter, which **TWO** structures will it pass through next (**in sequence**)?
3. _____ Carries blood high in metabolic wastes to the kidney.
4. _____ Returns filtered blood to body circulation.
5. _____ When this structure is full, nerve endings signal the urge to urinate.
6. _____ Urine formation begins in this region of the kidney.
7. _____ Either urine or semen could exit the body through this tube.
8. _____ Plays a vital role in maintaining homeostasis in the body.
9. _____ This portion of the kidney funnels urine to the ureter.

**Check your answers and Figure 12-3 labels with your instructor before you continue.**

Exercise 12 ■ *Organ Systems in the Abdominal Cavity*

# ACTIVITY 3      INTERNAL STRUCTURE OF THE STOMACH

1. With your fingers, locate the **pyloric sphincter** of the fetal pig. It will feel like a **small, hard mass** in the stomach wall. Using the same method, locate the **cardiac (gastroesophageal) sphincter**.

   Referring to **Figure 12-4**, use your **scissors** to remove the entire stomach, being careful to include **both sphincters**.

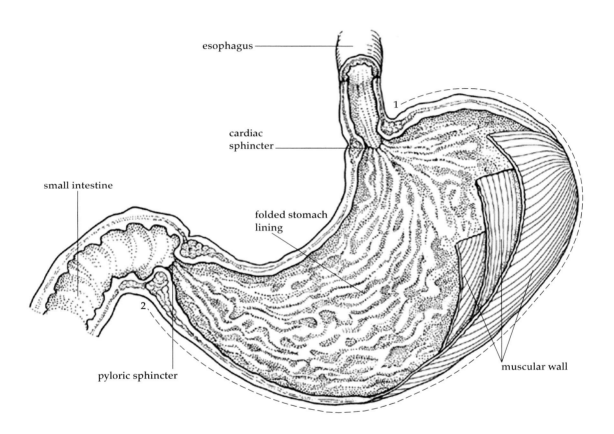

**FIGURE 12-4.** Internal Structure of the Stomach

2. Following the **dotted line** in **Figure 12-4**, cut from **number 1 to number 2** and open up the stomach.

   **Rinse** the inside of the stomach and **blot it dry** with paper towels.

3. Examine the **interior** of the stomach. **Observe** that the stomach wall is folded, like an accordion.

   Imagine that, disregarding what you know about dietary guidelines, you stuff yourself with two cheeseburgers, a supersize order of fries, an apple pie, and a chocolate shake.

   **During this intake**, what happens to the **size** of your stomach? _____

   How are the **folds in the stomach wall** related to this change in size?

4. A sphincter is a **ring of muscle** that surrounds the opening of a hollow organ. When the sphincter muscles **contract**, the opening is shut.

   **Visualize** what happens when you pull the ties on a plastic trash bag and the bag closes. This is similar to the action of a sphincter muscle.

   The stomach has **two sphincters**. One is between the esophagus and the stomach (the **cardiac sphincter**), and the other is between the stomach and the small intestine (the **pyloric sphincter**).

   **(Circle one answer.)** If there is food in the esophagus, the cardiac sphincter will be **opened/closed**.

   What is happening to the food in the **stomach** while **both sphincters are closed**?

   If the **pyloric sphincter is open**, where is the food going? _____

**Exercise 12** ■ *Organ Systems in the Abdominal Cavity*

5. With one hand, feel the pyloric sphincter. At the **same time**, use the other hand to feel the cardiac sphincter.

   How are the two sphincters **different**?

   Which sphincter do you think is **stronger**? _____

   The **cardiac sphincter** prevents food in the stomach from backing up into the esophagus.

   The **pyloric sphincter** controls the **rate** at which food enters the small intestine. No matter how full your stomach is, food is always released **gradually** into the small intestine.

✓ Comprehension Check

1. Considering your examination of the stomach, **explain** what happens during **vomiting**.

2. What would happen if your **pyloric sphincter** were not able to **close**?

3. What would happen if your **pyloric sphincter** were not able to **open**?

4. What would happen if the **cardiac sphincter** didn't close completely and stomach juices were allowed to enter the esophagus?

## ACTIVITY 4 — EXAMINING THE SMALL INTESTINE

1. Hold up a loop of the small intestine and **examine the mesenteries**.

   In addition to holding the small intestine in position, the mesenteries provide a passageway for **large numbers of blood vessels**.

2. Using your **scissors**, carefully **snip through the mesenteries** and start separating the loops of the small intestine.

   **Remove the small intestine.**

3. Lay the intestine out on the laboratory counter and use a **meter stick or yardstick** to measure the length.

   The length of my pig's small intestine is _____.

   **(Circle one answer.)** The small intestine is **shorter/longer** than I expected.

   If the small intestine is longer than you expected, you might be interested in the origin of the name "small" intestine. The intestine is named for its **small diameter (and not its length)**.

   The large intestine has twice the diameter of the small intestine **(but only half the length)**.

4. Although each person is slightly different, the human small intestine averages about **20 feet (6 meters)** in length.

**Exercise 12** ■ *Organ Systems in the Abdominal Cavity*

Using your knowledge of the digestive system, **explain** how increasing the length of the small intestine would help this organ **perform its functions**.

5. Remove approximately **two inches (five cm)** from the small intestine. Using your **scissors**, cut **open** the piece of intestine **lengthwise** and place it on the stage of the **dissecting microscope**.

6. Observe that the inside of the small intestine is **not smooth**.

    The interior wall is **folded**, and **each fold is carpeted** with slender fingerlike projections called **villi** (singular is **villus**).

7. Although you cannot see them at this magnification, each of the villi is covered, in turn, with even more tiny projections called **microvilli** (see **Figure 12-5**).

    Located inside each of the villi is a network of capillaries and lymphatic vessels (**lacteals**). These vessels pick up nutrients as they are absorbed, so that they can be distributed to all parts of the body.

    The small intestine is quite a narrow tube. However, if the villi and the microvilli in the intestinal tract were laid out flat, the surface area would be **half the size of a basketball court**!

8. After you have completed your dissection, do the following:

    a. Remove the strings from the limbs.

    b. Your instructor will tell you how to dispose of your pig and any remaining tissues.

    c. Wash and dry all instruments and trays.

**FIGURE 12-5.** Interior of the Small Intestine

# ACTIVITY 5 — THE IMPORTANCE OF SURFACE AREA

1. **Why does the small intestine need such a large surface area?** The need to increase surface area is common throughout the plant and animal kingdoms. On the laboratory counter, you will find a demonstration of one method used to increase surface area in a plant root (**a radish seedling**).

   **(Circle one answer.)** The radish root tip appears **smooth/fuzzy**.

   What **two** important substances enter a plant through the **roots**?

   _____     _____

2. The many slender extensions you can see covering the root of the radish seedling are called **root hairs**.

   Use **Figure 12-6** to determine how **root hairs** help plant roots function.

   In **both** the pictures in **Figure 12-6, the dark line represents the outside of the root surface.**

   Following the dark line, **place an X in every square that the line passes through** (any square that touches the exterior of the root surface). Place only one X in each square even if the line passes through the square twice. A few Xs have been drawn into the figure to get you started.

Without Root Hairs             With Root Hairs

**FIGURE 12-6.** Comparison of Surface Area with and without Root Hairs

3. For each picture in **Figure 12-6**, **count the number of squares with an X** and enter your results below:

   **Without root hairs** _____     **With root hairs** _____

4. Imagine that each graph square represents **one drop of water**. Which root (with or without root hairs) can absorb the **greatest number of water droplets at one time**? **Explain** your answer.

## Comprehension Check

1. Imagine that each graph square represents **one nutrient molecule**.

   Which type of intestinal lining (with or without villi) can absorb the **greatest number of nutrient molecules at one time? Explain your answer.**

2. Considering your answer to the above question, how are **microvilli** beneficial for small intestine function?

**Check your answers with your instructor before you continue.**

Exercise 12 ■ *Organ Systems in the Abdominal Cavity*

## ACTIVITY 6 — A CLOSER LOOK AT THE INSIDE OF THE SMALL INTESTINE

1. Referring to **Figure 12-7**, locate the **villi** (projecting toward the center of the tube).

2. At the base of the villi, you will see many **small**, circular structures.

   These are the **intestinal glands**. Intestinal glands produce **digestive enzymes** that perform the final breakdown of food molecules.

   Add a **label** for **intestinal glands** to the diagram in **Figure 12-7**.

3. Opposite the intestinal glands, you will see a cluster of **larger** circles.

   These are **lymphoid nodules,** small lymph nodes that act as **bacterial filters** in the digestive, respiratory, and urinary tracts.

   Which other body locations contain **lymph nodes**?




   Add a **label** for **lymphoid nodules** to the diagram in **Figure 12-7**.

4. You can see **two layers of muscle tissue** around the outside of the intestine.

   The **innermost** layer of muscle fibers runs in a circle around the tube and is referred to as **circular muscle**.

   The **outermost** layer has the muscle fibers arranged along the **length** of the tube and is called **longitudinal muscle**.

   Add **labels** for the **circular and longitudinal muscle** layers to the diagram in **Figure 12-7**.

5. The **circular and longitudinal** muscle layers **alternately** contract, producing mixing movements and moving food through the small intestine with waves of **peristalsis**.

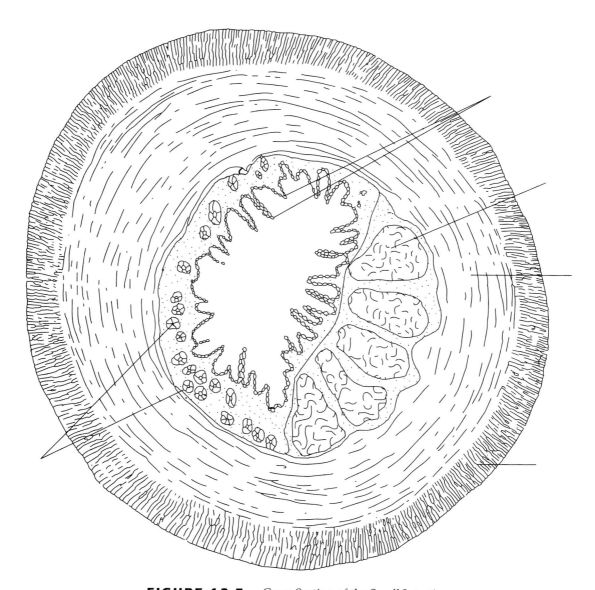

**FIGURE 12-7.** Cross Section of the Small Intestine

 **Check your figure labels with your instructor before you continue.**

**Exercise 12** ■ *Organ Systems in the Abdominal Cavity*

# ACTIVITY 7 — EARTHWORM LOCOMOTION

Preparation

To see waves of **peristaltic action,** we will examine locomotion in an earthworm.

Earthworm movement is a result of alternating muscle layers in the body wall. Alternating contractions of the **circular and longitudinal muscle** layers allow the earthworm to move forward. Segments may also move in the reverse direction, allowing the earthworm to move backward.

1. Work in groups. **Get an earthworm** and **several clean, dampened paper towels**.

> **Attention:**
> Wash your hands before touching the worms!

2. **Lay out** the dampened paper towels on your laboratory counter and **place the earthworm** on the towels.

    Observe the **muscle contractions** as the earthworm moves forward. From your observations, **describe how the earthworm's body changes** as it moves.

3. For a closer look at the alternating action of circular and longitudinal muscles, place the earthworm on the stage of your **dissecting microscope**. Similar muscle contractions in the small intestine mix and move food.

    When you have completed your observations, **return the earthworm** to the container on the supply table.

4. How is the **process of peristalsis** related to earthworm locomotion? (If you need more information to answer this question, refer to your textbook.)

5. **(Circle one answer.)** The circular and longitudinal muscles in the small intestine are composed of **smooth muscle tissue/skeletal muscle tissue/cardiac muscle tissue**.

Self Test

Fill in the blanks with the **most appropriate** answer. Answers can be used **only once**.

a. Lymphoid nodules
b. Root hairs
c. Villi
d. Circular muscle
e. Longitudinal muscle
f. Peristalsis
g. Surface area
h. Mesenteries

1. _____ Membranes that hold the small intestines in position.

2. _____ Small projections that increase surface area in the small intestine.

3. _____ Involuntary muscle running **around** the intestine.

4. _____ Contractions that move food through the digestive tract.

5. _____ The **outermost** muscle layer in the small intestine.

6. _____ Has a **function similar** to the villi of the small intestine.

7. Which of the following organ systems are represented in the **abdominal cavity**? Circle ALL correct answers and **give an example of an abdominal organ or structure** that belongs to each system.

| ORGAN SYSTEM | EXAMPLES |
|---|---|
| a. Digestive system | |
| b. Respiratory system | |
| c. Urinary system | |
| d. Circulatory system | |

**Exercise 12** ■ *Organ Systems in the Abdominal Cavity*

8. When you experience **heartburn,** which **stomach** structure is not functioning correctly? **Explain your answer.**

9. **Bulimia** is an eating disorder that involves forced vomiting. Over a period of years, bulimics gradually lose the enamel coating from their teeth. Using your knowledge of the digestive system, explain what causes the loss of tooth enamel.

10. **Urinary incontinence** (inability to control urine flow) frequently occurs in the elderly. Which organ of the urinary system may be malfunctioning? Which structure is malfunctioning in that organ you named? **Explain your answer.**

11. **Figure 12-8** contains two designs for air sacs in a human lung. Which will be the **most efficient** in taking in oxygen and removing carbon dioxide? **Explain your answer**.

Air Sac Design A

Air Sac Design B

**FIGURE 12-8.** Comparison of Air Sac Designs

**Exercise 12** ■ *Organ Systems in the Abdominal Cavity*

Complete the crossword puzzle in Figure 12-9 to review the pig structures and functions.

ACROSS
1. Pair of blood vessels that carry oxygen-poor blood that is high in wastes from the fetus to the placenta; called the umbilical _____
2. Most chemical digestion occurs here; enzymes break down proteins, fats, carbohydrates, and nucleic acids
3. External genital structure in female pigs
4. Part of the immune system; site of T-cell development
5. Muscular tube that connects the mouth to the stomach
6. Part of the immune system; blood-filtering organ and storage site for red blood cells and some white blood cells
7. Thin connective tissue membranes that hold the internal organs in place and provide a passageway for blood vessels and nerves that supply the organs
8. Bony roof of the mouth
9. Houses the vocal cords, the vibrations of which make speech possible; composed of cartilage
10. Fleshy sac that contains the testes in males
11. Endocrine (hormone-producing) gland; produces a hormone that helps in regulating the metabolic activity of the body
12. Secretes digestive enzymes into the small intestine; also functions as an endocrine organ, producing two hormones that control blood sugar levels
13. Has many functions including storage of energy reserves, detoxification of poisons, and bile formation

DOWN
1. Stores bile formed by the liver; releases it through the bile duct into the small intestine (for fat digestion)
2. Baglike organ that stores food and releases it gradually into the small intestine; mechanical and some chemical digestion occurs here
3. Air channel to the lungs; surrounded by tough, elastic rings of cartilage
4. Excretory organ that stores urine
5. Hood-shaped flap of tissue that covers the opening to the trachea; prevents food and liquid from entering the air passageways
6. Composed of millions of tiny sacs, which are the site of gas exchange for the whole body; oxygen enters and carbon dioxide is removed
7. Paired blood vessels that drain deoxygenated blood from the head region back to the heart; called the jugular _____
8. Excretory organ that removes nitrogen-containing wastes and maintains the proper solute concentration of body fluids
9. Digestive organ whose main function is the reabsorption of water from wastes; prevents dehydration
10. Muscular pump consisting of four chambers; pumps blood to the lungs and the body tissues

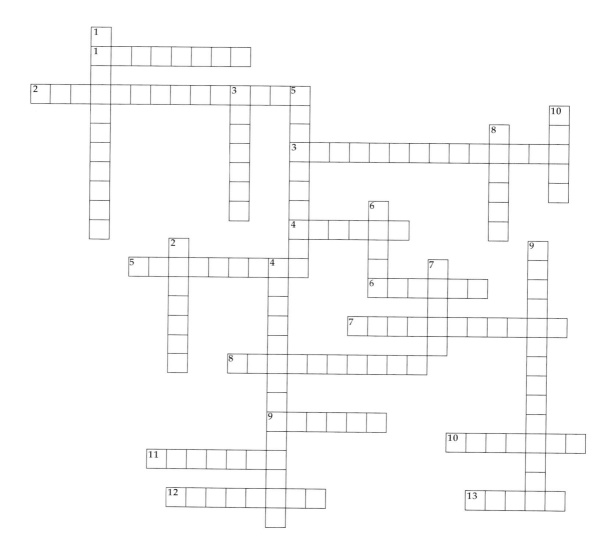

**FIGURE 12-9.** Crossword Puzzle

EXERCISE 13

# Nutrient Analysis of Foods and Choices for Good Health

## Objectives
After completing this exercise, you should be able to:
- test an unknown sample for the presence of sugars, starch, protein, and lipids using indicator chemicals
- relate the nutrient content of a food to its original function in plants and animals
- discuss the benefits and drawbacks of using carbohydrates versus lipids as energy storage molecules for embryos
- use the Federal Dietary Guidelines to analyze the nutritional value of your meals.

## CONTENT FOCUS

All living organisms are composed of various types of **organic molecules** such as carbohydrates, lipids, proteins, and nucleic acids that make up their body tissues. Each of these four types of organic molecules (carbon-based compounds) has different functions in the body. The foods we eat are obtained from plants and animals. These foods are needed not only as a source of energy but also because they provide many essential nutrients. Just as different parts of your body have different structures to support their functions, the same is true of other animals and plants. The **types** of **organic chemicals present** in different body tissues are related to the **function** of those tissues.

In this exercise, you will be analyzing various food samples to determine which organic compounds are present. Carbohydrates, lipids, proteins, and nucleic acids **cannot be seen directly. Indicator chemicals** can show you if these organic compounds are present in a food sample by **changing color**. Each type of indicator chemical is specific for one type of organic compound.

Analyzing foods for organic compounds using indicator chemicals can present some problems. Often, the **food itself has colors** that may interfere with the test results. It is easier to determine if an appropriate color change has occurred if you have a **standard for comparison**. Therefore, we must first establish a set of **positive and negative standards (controls)** to be used for comparison in later experiments.

# ACTIVITY 1    POSITIVE AND NEGATIVE INDICATOR TESTS

1. Work in groups. Get the following supplies: **a ruler, a marking pencil, six test tubes, a test tube rack, a test tube holder, and a large beaker**.

2. Using the ruler, place a **line** on each test tube **1 cm from the bottom** of the tube. **Number each test tube 1 through 6**. Place the marked test tubes in the test tube rack.

*Benedict's Test for Simple Sugars*

1. The chemical indicator in Benedict's solution reacts with monosaccharides and some disaccharides, but **only when the indicator solution is heated**.

    Place the large beaker, half filled with tap water, on a hot plate. Set the temperature control to high until the water boils.

2. Fill **test tube #1** to the line with **glucose** solution. Fill **test tube #2** to the line with **distilled water**.

3. Fill the dropper with Benedict's solution. What color is Benedict's solution? _____

4. To **test tubes #1 and #2**, add **one dropper full of Benedict's solution. Shake** the test tubes to mix.

5. **Using a test tube holder, carefully** place the tubes into the boiling water for two to five minutes.

> ### Caution!
> **Hot glass looks exactly like cold glass! Use the test tube holder to remove test tubes from the boiling water!**
>
> **Don't leave the test tube holder clamped on the test tube while in the boiling water. Hot metal looks exactly like cold metal!**

**Exercise 13** ■ *Nutrient Analysis of Foods and Choices for Good Health* **235**

6. **Using the test tube holder** remove the tubes and **observe** the color in each test tube.

    If you see a color change ranging from **green through yellow, orange, or red**, this is a positive test for simple sugars.

    If the color in the tube has **not changed**, this is a **negative** test for simple sugars.

> **Note:**
> Save tubes #1 and #2 for use in later experiments.

7. What happened to the **color** in **tube #1**? _____

8. What happened to the **color** in **tube #2**? _____

### ✓ Comprehension Check

1. Which tube showed a positive test for glucose? _____

2. On the basis of the **results of Benedict's test, glucose** could be correctly identified as belonging to which of the following groups?

    a. Lipids
    b. Starches
    c. Proteins
    d. Simple sugars
    e. Both b and d are correct

Iodine Test for Starch

1. Fill **test tube #3** to the line with **starch** solution. Fill **test tube #4** to the line with **distilled water**.

2. Fill the **dropper with iodine**. What color is iodine? _____

3. To **both tubes**, add one **dropper full of iodine. Shake** the test tubes to mix.

    **Observe** the color in each test tube.

    What color is the liquid in **test tube #3**? _____

    What color is the liquid in **test tube #4**? _____

> **Note:**
> Save tubes #3 and #4 for use in later experiments.

4. Which tube showed a positive test for starch? _____

5. Place **a small drop of iodine** in the box.

On the basis of the results of the iodine test, what can you **conclude about the paper** in the laboratory manual?

Biuret Test for Protein

1. Fill **test tube #5** to the line with **albumin (egg white)** solution. Fill **test tube #6** to the line with distilled **water**.

2. Fill the dropper with **Biuret solution**. What color is Biuret solution? _____

3. To **tubes #5 and #6**, add one dropper full of **Biuret solution. Shake** the tubes to mix.

4. **Observe** the color in each test tube.

> **Note:**
> Save tubes #5 and #6 for use in later experiments.

**Exercise 13** ■ *Nutrient Analysis of Foods and Choices for Good Health*

5. What happened to the color in **tube #5?**

6. What happened to the color in **tube #6?**

7. Which tube showed a **positive** test for **protein?** _____

Sudan IV Test for Lipids

1. Get the following supplies: **one piece of filter paper, forceps, a petri dish, and dropper bottles of vegetable oil, distilled water, and Sudan IV stain.**

2. Place the filter paper on a **clean** piece of paper (**not** on the laboratory counter, which may be dirty).

   Using a **pencil**, draw **two** circles **the size of a dime**, spaced **apart** on the filter paper. Label the first circle **"oil"** and the second circle **"water."**

3. Place **one drop of oil** into the circle labeled "oil" and **one drop of distilled water** into the circle labeled **"water."**

4. **Set the filter paper aside to dry**. If needed, use a hair dryer **on low power** to evaporate excess liquid.

5. When the filter paper is **completely dry**, place it in the **petri dish** and cover the paper with **Sudan IV solution**. Let the paper soak in the stain for **three minutes**.

6. While the paper is soaking, get a **glass bowl** from the **supply area** and **fill it** with **distilled water**.

7. When the three minutes are up, place the filter paper into the bowl of distilled water. **Rinse gently** for **one minute**.

   Remove the filter paper from the water with the forceps and **observe the color of the two circles**.

8. **A dark red spot** indicates a **positive** test for lipids.

   A **pale pink** color, **no different from the rest of the paper**, should be considered **negative**.

9. **Record** the results of your test (positive or negative) below. Enter a "+" if the test was **positive** and a "−" if the test was **negative**.

   "oil" circle _____ "water" circle _____

> **Note:**
> **Save your test paper for use in later experiments.**

## ✓ Comprehension Check

Using the results from the tests you have just completed, fill in **Table 13-1**.

| TABLE 13-1 INDICATOR TESTS | | | |
|---|---|---|---|
| Indicator Test | Tests for | Negative Result (Color) | Positive Result (Color) |
|  |  |  | green through red |
|  | starch |  |  |
| Sudan IV Test |  |  |  |
| Biuret Test |  | light blue |  |

**Check your answers with your instructor before you continue.**

# ACTIVITY 2     TESTING FOOD SAMPLES

*Preparation*

Have you ever wondered what is really in the foods you choose? Is a food high in fat or protein? Is it a good source of carbohydrates? The answers to these questions become clearer when food samples are analyzed to determine which organic compounds are present.

In the next activity, you will analyze several commonly eaten foods and determine their nutrient content. Solid food samples have been blended into liquid for easier testing.

1. For each of the following foods, form a **hypothesis** about which organic compounds it will contain.

   **Record** your hypotheses in **Table 13-2**.

   For each hypothesis, enter a "+" if you think the test will be **positive** and a "−" if you think the test will be **negative**.

2. Work in groups. Your instructor will assign several foods for your group to analyze. Your group will **share its data** with other groups that are analyzing different foods.

Exercise 13 ■ *Nutrient Analysis of Foods and Choices for Good Health*

### TABLE 13-2
### HYPOTHESES FOR FOOD EXPERIMENTS

| Food | Simple Sugars | Starch | Protein | Lipid |
|---|---|---|---|---|
| Peanut butter | | | | |
| Tuna | | | | |
| Refried beans | | | | |
| Milk | | | | |
| Hamburger | | | | |
| Lettuce | | | | |

3. For each food sample to be analyzed, get **three clean test tubes**. You will also need **one piece of filter paper**.

4. Following the **test procedures exactly as you did in Activity 1**, analyze each food sample for the presence of **simple sugars, starch, protein, and lipids**.

5. **Compare** your test results to the **controls** you **saved from Activity 1**.

6. **Record** the results of your tests (positive or negative) in **Table 13-3** and also on the **master chart at the front of the room**.

    Enter a "+" if the test was **positive** and a "−" if the test was **negative**.

7. **Complete Table 13-3** by entering the results from the **master chart** at the front of the room.

### TABLE 13-3
### RESULTS OF FOOD EXPERIMENTS

| Food | Simple Sugars | Starch | Protein | Lipid |
|---|---|---|---|---|
| Peanut butter | | | | |
| Tuna | | | | |
| Refried beans | | | | |
| Milk | | | | |
| Hamburger | | | | |
| Lettuce | | | | |

## Comprehension Check

1. List the foods tested that contain **all three** of the following: **carbohydrates, proteins, and lipids**.

   It is possible to relate the nutrient content of a food to its original function in plants and animals. Using the information provided below, **explain** why certain nutrients occur in high levels in particular foods.

2. This food comes from the leaf of a plant. Leaf cells perform photosynthesis. Which of the six foods is being described? _____

   **(Circle one answer.)** If photosynthesis is taking place, I would predict a positive indicator test for **simple sugar/lipid/protein**.

3. Of the foods tested, which had high levels of **lipids**? _____

4. Of the foods tested, which use **lipids** as a high energy source for young animals? _____

5. **Muscle tissue** is composed of **interwoven protein fibers**. Using this information, **name two foods** made of **muscle** tissue that gave you a **positive** test for **protein**.

6. The hamburger and tuna tested **negative** for starch. **Explain** why foods of this type would **not** have starch present.

**Check your answers with your instructor before you continue.**

# ACTIVITY 3 — AN EMBRYO IN A PEANUT SEED

Preparation

A seed contains a **plant embryo** packaged with its **food supply**. The embryo and stored food are protected by a tough seed coat. A developing embryo has the beginnings of a stem, tiny leaves, and a root. You can see the tiny stem, leaf, and root structures of an **embryonic peanut plant** if you carefully pull the two peanut halves apart.

The nutrients in seeds are not only used by germinating plant embryos are also consumed by animals, including humans.

1. In your classroom, you will find a demonstration of a **separated peanut seed** that can be viewed with a dissecting microscope.

2. Look for the following structures and label them on the diagram in **Figure 13-1**: **root, stem, leaves,** and **stored food supply**.

**FIGURE 13-1.** Peanut Seed

## Comprehension Check

1. Food may be stored for the use of an adult plant or as energy for the development of an embryo. The stored food is packaged with an embryo into a structure called a **seed**.

   List **three commonly eaten foods** produced by plants or animals to provide stored energy for embryos:

   _____    _____    _____

2. Of the foods tested in Activity 2, which use **lipids** as an energy source for a **developing embryo?**

3. **(Circle all correct answers.)** The following probably have high lipid levels:

   a. Chicken egg      c. Sunflower seed      e. Coconut
   b. Carrot           d. Spinach             f. Almond

   **Explain your answer.**

**Check your answers with your instructor before you continue.**

# ACTIVITY 4 — DIETARY INTAKE AND GOOD HEALTH

Preparation

When you eat a meal, you are probably consuming a combination of carbohydrates, proteins, and lipids. Imagine this situation. You've been stuck in classes all morning (just grabbed a bag of peanuts from the candy machine), worked in the afternoon, picked up the baby-sitter, and now you're late for a meeting of your study group. Of course, you're starving! On the way to the library, you stop off at your favorite fast food restaurant to refuel. You order a double hamburger with cheese, a large order of fries, and a chocolate milk shake.

One person in your study group is taking a nutrition course and has been studying dietary analysis. He was shocked to discover that his normal intake was not even close to the guidelines for a healthy diet. As he tells you about this, you wonder whether your diet is any better. Below, you will find the instructions provided by your friend for calculating the percentage of carbohydrates, protein, and fat in your diet. He also thoughtfully provided you with a copy of the government recommendations for a healthy diet (**Table 13-4**). **Table 13-5** summarizes your meals.

## TABLE 13-4
### RECOMMENDED DIETARY GUIDELINES

| NUTRIENT | RECOMMENDED PERCENTAGE OF DAILY CALORIC INTAKE | |
|---|---|---|
| Carbohydrates | 55–60% | |
| Protein | 10–15% | |
| Fat | no more than 30% | |
| 1 g carbohydrate contains 4 kcal | 1 g protein contains 4 kcal | 1 g fat contains 9 kcal |

**Follow the instructions below to calculate the percentage of carbohydrate, protein, and fat you consumed today.**

1. Begin your calculations with the hamburger. **Multiply** the grams of carbohydrates in the hamburger by **4** (since there are 4 kcal per gram of carbohydrate).

   **Record the answer in Table 13-5**. Perform the same calculations for the **protein and fat** in the hamburger (using the appropriate number of kilocalories) and record your answers.

2. Calculate the **kilocalories** for the other foods listed in **Table 13-5** and record your answers.

3. Total each column in **Table 13-5** and **record** your answers.

   Add the total **kilocalories** from **carbohydrates, protein, and fat** and record the answer at the bottom of **Table 13-5**.

### TABLE 13-5
### DIETARY CALCULATIONS

| Food Eaten | Carbohydrate (g) | Carbohydrate (kcal) | Protein (g) | Protein (kcal) | Fat (g) | Fat (kcal) |
|---|---|---|---|---|---|---|
| Double beef hamburger with cheese | | 54 | 51 | | 60 | |
| Fries (large order) | | 43 | 5 | | 20 | |
| Chocolate milk shake | | 49 | 9 | | 10 | |
| Bag of peanuts (2 ounces) | | 10 | 14 | | 28 | |
| | | | | | | |
| Column Totals | | | | | | |
| Total kcal: | | | | | | |

4. Calculate the **percent of the day's diet that came from carbohydrates, protein, and fat**.

   ____ % of carbohydrates in diet = $\dfrac{\text{total kcal from carbohydrates}}{\text{total kcal consumed}} \times 100 =$ ____ %

   ____ % of protein in diet = $\dfrac{\text{total kcal from protein}}{\text{total kcal consumed}} \times 100 =$ ____ %

   ____ % of fat in diet = $\dfrac{\text{total kcal from fat}}{\text{total kcal consumed}} \times 100 =$ ____ %

   $\overline{\qquad 100 \ \%}$

5. Compare your calculated percentages to the federal government's recommended values.

   Was your intake today in line with recommended dietary goals? _____

   If not, which nutrients did not match the guidelines?

**Exercise 13** ■ *Nutrient Analysis of Foods and Choices for Good Health* **245**

6. Place an X in front of any of the following that are good suggestions for improving your day's food intake.

   ___ Bring an apple or an orange for a snack instead of the peanuts.

   ___ Replace the fries with a baked potato.

   ___ Replace the hamburger with pizza.

   ___ Replace the milk shake with low-fat milk or juice.

   ___ Eat a candy bar instead of the peanuts.

7. If your body requires **1500 kcal per day**, how many excess kilocalories did you eat in the day described in **Table 13-5**? _____

   If you consumed the same amount of excess kilocalories each day for the last three weeks, **how many excess kilocalories** did you accumulate? _____

   Every time you accumulate 3500 excess kilocalories, you gain a pound.

   How many pounds have you gained over the last three weeks? _____

8. You want to lose this excess weight, but with your schedule, you're too busy to exercise regularly. However, if you park in the furthest lot from the building, you can get to class in 15 minutes of quick walking. Every time you do this, you will burn **150 kcal**.

   If you walk quickly to and from the parking lot once a day, you can lose the weight you gained in only _____ days.

■ **Check your answers with your instructor before you continue.**

## ACTIVITY 5                                          READING FOOD LABELS

*Preparation*

You are sitting at breakfast with your bowl of cereal or toast and notice that nutritional information is included on the package label. This information is mandated by the U.S. Food and Drug Administration (FDA). The food label is a valuable tool for evaluating packaged food products. You can compare the information on a food label with the federal dietary guidelines you learned about in the previous activity to make diet selections that will recommended for good health.

Read the food label in **Figure 13-2** and answer the following questions.

This section lists the number of calories in one serving and also the number of fat calories per serving.

**Nutrition Facts**
Serving Size 1 cup (228g)
Servings Per Container 2

**Amount Per Serving**
**Calories** 260   Calories from Fat 120

| | % Daily Value* |
|---|---|
| **Total Fat** 13g | 20% |
| Saturated Fat 5g | 25% |
| **Cholesterol** 30mg | 10% |
| **Sodium** 660mg | 28% |
| **Total Carbohydrate** 31g | 10% |
| Dietary Fiber 0g | 0% |
| Sugars 5g | |
| **Protein** 5g | |

| Vitamin A 4% | • | Vitamin C 2% |
|---|---|---|
| Calcium 15% | • | Iron 4% |

* Percent Daily Values are based on a 2,000 calorie diet. Your daily values may be higher or lower depending on your calorie needs:

| | Calories: | 2,000 | 2,500 |
|---|---|---|---|
| Total Fat | Less than | 65g | 80g |
| Sat Fat | Less than | 20g | 25g |
| Cholesterol | Less than | 300mg | 300mg |
| Sodium | Less than | 2,400mg | 2,400mg |
| Total Carbohydrate | | 300g | 375g |
| Dietary Fiber | | 25g | 30g |

Calories per gram:
Fat 9 • Carbohydrate 4 • Protein 4

INGREDIENTS: WATER, ENRICHED MACARONI (ENRICHED FLOUR - NIACIN, IRON, THIAMINE, MONONITRATE, AND RIBOFLAVIN), EGG WHITE, FLOUR, CHEDDAR CHEESE (MILK, CHEESE CULTURE, SALT, ENZYME), SPICES, MARGARINE (PARTIALLY HYDROGENATED SOYBEAN OIL, WATER, SOY LECITHIN, MONO AND DIGLYCERIDES, BETA CAROTENE FOR COLOR, VITAMIN A PALMITATE), AND MALTODEXTRIN

All the information on a nutritional label is per serving; however, many packages contain multiple servings.

Use the percent daily value to easily compare products and to quickly tell if a serving of food is high or low in nutrients.

The listed nutrients are those considered to be most important for good health.

Ingredients are listed in descending order of weight. The sources of some ingredients, such as certain flavorings, are stated by name to make it easier for people to identify ingredients that they wish to avoid for health, religious, or other reasons.

**FIGURE 13-2.** Sample Food Label

**Exercise 13** ■ *Nutrient Analysis of Foods and Choices for Good Health*

1. If you ate **all the food** in the labeled container, how much of each of the following would you have eaten? Enter your calculations in **Table 13-6** below.

| TABLE 13-6 FOOD LABEL CALCULATIONS | | | |
|---|---|---|---|
| Total Fat | | Total Carbohydrate | |
| Saturated fat | | Dietary fiber | |
| Cholesterol | | Sugar | |
| Sodium | | Protein | |

2. Calculate the fat calories recommended each day if you were on a **2000 calorie diet** (remember, the maximum should be 30% of your calories from fat).

    Recommended calories from fat:

    $2000 \times .30 =$

3. Calculate the percentage of "allowed" dietary calories from fat you would consume if you ate the entire container of macaroni and cheese discussed above.

    Percentage of calories from fat eaten:

    $$\frac{\text{fat calories eaten}}{\text{recommended calories from fat}} =$$

4. Would you consider this food a good choice if you were trying to control the amount of fat in your diet? **Explain your answer**.

5. If the number of calories from fat (listed in the upper right hand corner of the food label) was missing, how would you determine the approximate number of fat calories per serving?

Self Test

1. You add **Biuret solution** to your morning orange juice. The Biuret solution does not change color. What can you **conclude** from this experiment?

2. What do you think would happen if you placed a drop of iodine on your baked potato at dinner?

On your steak?

**Explain your answer.**

**Exercise 13** ■ *Nutrient Analysis of Foods and Choices for Good Health* **249**

3. Match the foods in **Table 13-7** with their functions in the living organism they come from. For example, why does a cow produce milk? (Hint: Not for your morning cereal!)

| TABLE 13-7 MATCHING FOODS AND THEIR FUNCTIONS ||
|---|---|
| Food | Function in Plant or Animal |
| Peanut butter | |
| Tuna | |
| Bean | |
| Milk | |
| Hamburger meat | |
| Lettuce | |

Use the food label below to answer the following questions:

**Nutrition Facts**
Serving Size 1 oz. (28g)
Servings Per Container about 4

**Amount Per Serving**
**Calories** 125   Calories from Fat 40.5

| | % Daily Value* |
|---|---|
| **Total Fat** 4.5g | 7% |
| Saturated Fat 0g | 0% |
| Trans Fat 0g | 0% |
| **Cholesterol** 0mg | 0% |
| **Sodium** 70mg | 3% |
| **Total Carbohydrate** 18g | 18% |
| Dietary Fiber 2.25g | 8% |
| Sugars 0g | |
| **Protein** 2g | |

| Vitamin A 5% | • | Vitamin C 25% |
|---|---|---|
| Calcium 0% | • | Iron 4% |

4. How many milligrams (mg) of sodium are in one serving of this product? _____

5. How many calories would you consume if you ate the entire contents of this package? _____

6. How many grams of fat are in one serving of this product? _____

7. How many calories from protein are in three servings of this product? _____

8. How many servings of this product would you need to consume in order to obtain the minimum daily requirement of Vitamin C? _____

EXERCISE 14

# The Reproductive System

## Objectives
After completing this exercise, you should be able to:
- identify and explain the function of each of the major organs that make up the male reproductive system
- identify and explain the function of each of the major organs that make up the female reproductive system
- compare and contrast fetal pig reproductive structures with the comparable structures in humans
- trace the path of gametes through the male and female reproductive systems
- correlate sperm production with meiotic divisions
- identify the stages of follicular development within the ovary
- correlate the stages of follicular development to hormone production
- trace the stages of egg and follicle development through ovulation and corpus luteum formation.

## CONTENT FOCUS

Understanding the reproductive system is important when considering the many issues of reproductive health that face today's citizens. Issues such as endometriosis, sexually transmitted infections, and control of fertility are all better understood when you have a firm knowledge of reproductive anatomy and physiology.

The urinary and reproductive systems of mammals contain several shared structures, so, as part of your preparation for this dissection, you may wish to review the urinary structures (and their functions) that you examined in Exercise 12.

## ACTIVITY 1

## DISSECTION OF THE PIG REPRODUCTIVE SYSTEM

Although you are looking at the reproductive organs of male and female fetal pigs, most of the structures and their functions are similar to those found in the human reproductive system.

1. Get your pig from the storage container and set it up on a tray as you did in previous dissections.

   Get a **set of dissecting instruments**.

2. **Based on the gender of your pig**, follow the appropriate instructions below.

   When you have completed your dissection, examine the organs and structures in a pig of the opposite gender.

*Dissection of a Male Fetal Pig*

1. Locate the urogenital opening, posterior to the umbilical cord. Pinch the skin around the opening. You should feel a muscular tube - the **penis**. You cannot see the **penis** on the exterior of the body since pigs, like many other four-legged animals, have a **retractable** penis.

2. Beginning at the urogenital opening, use your scissors to cut along the body midline **through the skin only** for a couple of inches.

   Using a blunt probe, locate the **urethra**, a white tube beneath the skin. The urethra **transports urine** out of the body through the **urogenital opening**. When the penis is inserted into the vagina, sperm pass through the urethra to the female reproductive tract.

   As you continue your dissection, refer to **Figure 14-1** to see the positions of the various organs.

   Continue your cut through the skin in a posterior direction to expose the entire length of the urethra. As you cut, **be careful not to disturb the testes**, which can be seen on either side of the urethra in the scrotal sac.

3. Using scissors, remove the **tissue between** the exposed urethra and the **urinary bladder** (which is located between the two umbilical arteries), but do not remove the bladder.

   The **urinary bladder** is a **temporary storage organ for urine**. When the bladder fills, nerve endings in its muscular walls are stimulated, causing the muscles to contract for urination.

# Exercise 14 ■ *The Reproductive System*

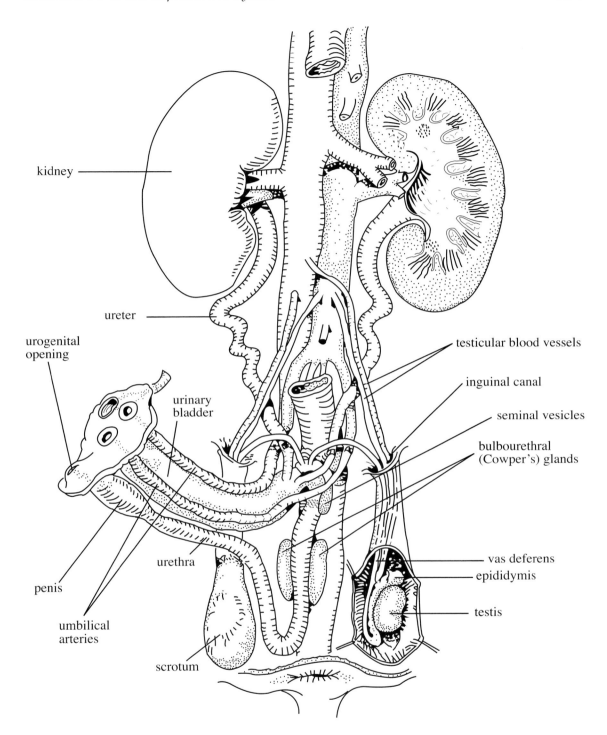

**FIGURE 14-1.** Urogenital System of the Male Pig

You can now see where the bladder and urethra are joined. Follow the urethra until it disappears beneath the **pubic symphysis** (midway between the two hipbones). The urethra will reappear in the region of the scrotum just ventral to the anus.

4. Slide the string out from beneath the tray **between the hind legs only**.

   Lift the posterior end of the pig. **While holding the posterior end elevated**, insert the scissors into the **anterior end** of the hipbones (the **pelvis**).

   **Keeping the tip of your scissors pointing up**, make a cut **about an inch** long through the pelvic bones.

   Bend the hind legs backwards until the pelvis cracks. The entire length of the urethra should now be visible (see **Figure 14-1**).

5. Carefully separate the urethra from the rectum by lifting and cutting below the urethra.

   This will expose the **three male accessory glands** (see **Figure 14-2**). During ejaculation, these glands release the **seminal fluid** which mixes with and transports the sperm.

   A pair of large glands, the **bulbourethral (Cowper's) gland**s, are visible on both sides of the urethra **near the anus**. These can be seen more clearly by lifting the urethra.

   Anterior to the bulbourethral glands, you will see a pair of small glands, the **seminal vesicles**, on the dorsal surface of the urethra.

   **Between the two seminal vesicles**, you will see an even smaller gland, the **prostate**. Considering the small size of the pigs, this gland may be difficult to see.

6. Return the urethra to its original position and look into the abdominal cavity. Toward the anterior end of the bladder, you can observe two small tubes (the **vas deferens**), which transport sperm from the testes to the urethra (see **Figure 14-1**).

   Trace the vas deferens posteriorly until it passes through an opening in the abdominal wall. The openings, called **inguinal canals**, are formed during the descent of the testes from the abdominal cavity to the **scrotum**. The scrotum is an external pouch of skin that contains the testes.

   If the inguinal canal is damaged or weakened, a loop of the intestine may penetrate and be pinched in the opening of the inguinal canal. This is called an inguinal hernia and must be surgically repaired.

   The **temperature of the testes is regulated** by the relative position of the scrotum to the body wall. In humans, normal sperm production requires a temperature about three degrees Centigrade lower than normal body temperature.

**Exercise 14** ■ *The Reproductive System*

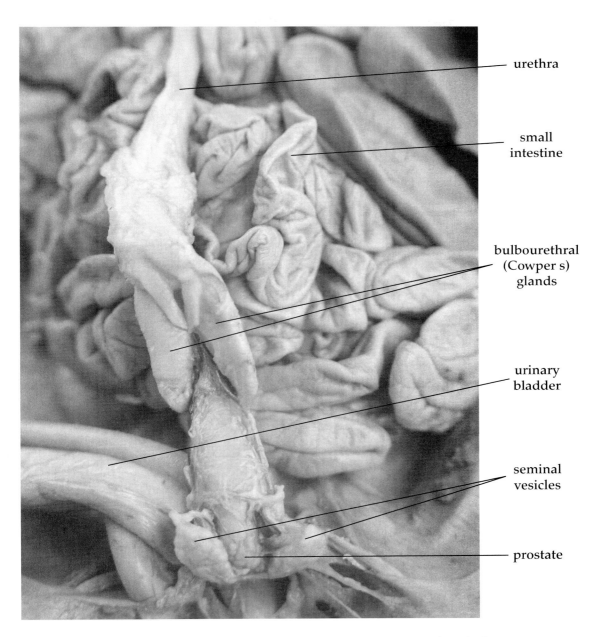

**FIGURE 14-2.** Male Accessory Glands (Dorsal View of Urethra)

Optimal sperm production requires a temperature that is a few degrees lower than normal body temperature. In cold temperatures, the testes are drawn closer to the body cavity and exposure to warmth reverses the process.

Pull gently on one vas deferens and note the movement of the attached **testis** within the **scrotum**.

7. Carefully remove the membranes covering one testis. This will expose several structures.

   The **testis** (where sperm are produced by meiosis) appears as an oval. Note that this region is well supplied with blood vessels.

   Sperm production occurs in the **seminiferous tubules**, microscopic structures within the testes. The testes also produce male sex hormones, such as testosterone.

8. Along one edge of the testis, is the **epididymis**, where sperm complete their maturation and are stored until ejaculation.

   Extending from the epididymis is a small tube—this is the **vas deferens** you were previously viewing from the abdominal cavity.

## ✓ Comprehension Check

1. In the correct order, list each organ and/or structure a sperm passes through from the site of production to the outside of the body.

2. Name **two** substances that travel through the **urethra**. What **two body systems** share this structure?

3. Sperm are stored in the _____ to complete their development.

Exercise 14 ■ *The Reproductive System*

4. In about 3% of male children, the testes fail to descend into the scrotum. Would this affect sperm production? **Explain** your answer.

5. If a male has a vasectomy (an operation that closes the vas deferens), will he be able to ejaculate normally? If so, where is the ejaculated fluid produced and what is missing from the ejaculate? **Explain** your answer.

**Check your answers with your instructor before you continue.**

Dissection of a Female Fetal Pig

1. If the small intestine has not been removed from the abdominal cavity, **retract it anteriorally** to expose the **ovaries** (small oval structures on each side of the rectum, posterior to the kidneys).

   Inside the ovary, are microscopic **follicles** that contain developing egg cells. The ovaries also produce the female sex hormones, estrogen and progesterone.

   As you continue your dissection, refer to **Figure 14-3** to see the positions of the various organs.

2. There are two tubes positioned close to the ovaries. The smaller tube is the **oviduct (fallopian tube)**, which **transports ovulated eggs** to the uterus and is the location where **fertilization** occurs.

   In the pig, the uterus is shaped like the letter **Y**. The larger, twisted tubes visible next to the ovaries are the **horns of the uterus**. In animals that have litters of young, uterine horns provide added space for multiple embryos to develop.

   The uterine horns join together into the **body of the uterus** (see **Figure 14-3**), close to the anterior end of the bladder. In humans, the body of the uterus is larger, and usually accommodates a single embryo.

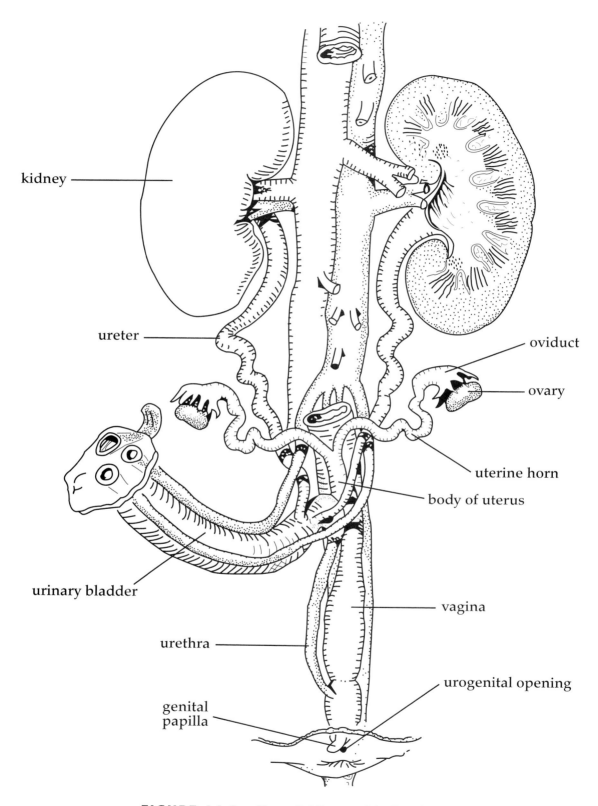

**FIGURE 14-3.** Urogenital System of the Female Pig

Exercise 14 ■ *The Reproductive System*

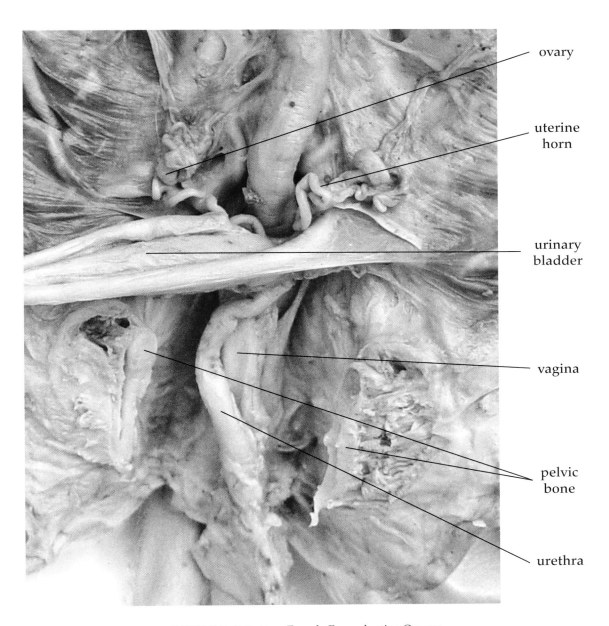

**FIGURE 14-4.** Female Reproductive Organs

3. Using scissors, separate the skin from the **urinary bladder** (which is located between the two umbilical arteries), by cutting posteriorally toward the anus, but do not remove the bladder.

   The **urinary bladder** is a **temporary storage organ for urine**. When the bladder fills, nerve endings in its muscular walls are stimulated, causing the muscles to contract for urination.

   You can now see where the bladder and urethra are joined. Follow the urethra until it disappears beneath the **public symphysis** (midway between the two hipbones).

4. Slide the string out from underneath the tray between the hind legs only.

   Lift the posterior end of the pig. **While holding the posterior end elevated**, insert the scissors into the **anterior end** of the hipbones (the **pelvis**).

   **Keeping the tip of your scissors pointing up**, make a cut **about an inch** long through the pelvic bones. Bend the hind legs backwards until the pelvis cracks. The urethra will be visible, but not completely exposed.

5. Cut through the skin and muscle on both sides of the urethra, proceeding posteriorally to the urogenital opening.

   You will see two parallel tubes. The more **ventral** of the two is the **urethra**. The urethra **transports urine** out of the body through the **urogenital opening**.

   **Dorsal to the urethra** you will see the **vagina** (see **Figure 14-4**). The vagina is the muscular tube into which **sperm are deposited** during sexual intercourse, and it also serves as the **birth canal**.

6. Insert the scissors between the vagina and the rectum and carefully cut in an anterior direction until the vagina separates from the rectum. Insert the scissors between the urethra and the vagina and carefully expose both tubes.

   The **urethra and vagina join** together to form a common passage that leads to the **urogenital opening**.

7. On the external surface of the pig surrounding the urogenital opening, you will see the **labia** (folds of skin). Gently spread the labia apart to see the tiny **clitoris** (erectile tissue).

**Exercise 14** ■ *The Reproductive System*

## ✓ Comprehension Check

1. In the correct order, list each organ and/or structure an egg will pass through from its site of production through fertilization, fetal development, and birth.

2. The female pig has a common opening for the urinary and reproductive systems. Do human females have the same anatomical pattern? If not, **explain the differences**.

3. If a woman has a tubal ligation (an operation that closes the oviducts), will she be able to ovulate normally? **Explain** your answer.

4. How does fertilization differ from implantation? Explain your answer, including the location where each takes place.

**Check your answers with your instructor before you continue.**

## ACTIVITY 2   DEVELOPMENT OF GAMETES IN THE TESTES

*Preparation*

In humans, sperm production begins at puberty and continues throughout a man's adult life. Sperm production (**spermatogenesis**) occurs by **meiosis** in the testes. Each ejaculation contains about 250 million sperm, so there is a continuous demand for large quantities of new sperm cells.

Sperm are formed within the seminiferous tubules of the testes. There are more than 100 meters of tightly coiled tubules in each testis. In the following activity, you will observe the development of sperm cells within the **seminiferous tubules** with the compound microscope.

1. Get a **compound microscope and a cross section slide of the testis**.

2. Observe the slide with the scanning lens (**4X**). You will see numerous circular structures. Each is a section through a **seminiferous tubule** (see **Figure 14-5**).

    Switch to **10X** and note that there are clusters of cells **between** the seminiferous tubules. These are referred to as **interstitial cells**. They produce **male sex hormones (androgens)**, the best known of which is **testosterone**.

3. Observe the slide on high power (**40X**) to get a closer view of the developing cells within **one tubule** (as shown in **Figure 14-6**).

    You will see that the tubule is packed with cells in various stages of development. Spermatogenesis begins at the outermost layer of the tubule.

    As repeated cell divisions occur, the developing cells move closer and closer to the center of the tubule. Therefore, your observations will also start near the outer wall of the tubule and move progressively inward toward the **lumen** (center) of the tubule (see **Figure 14-5**).

4. Begin your observations close to the outer wall of the tubule. You are looking for tiny cells that lie directly adjacent to the tubule wall. They are usually darker in color than other cells of the tubule. These are the **spermatagonia**, the stem cells that give rise to sperm.

    Refer to **Figure 14-6** for a correlation between the various stages of meiosis with the position of the developing sperm within the seminiferous tubules.

**Exercise 14** ■ *The Reproductive System* 263

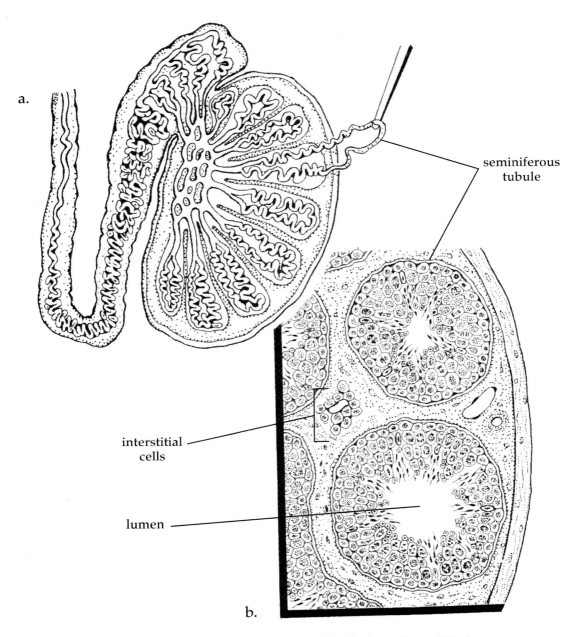

**FIGURE 14-5.** Spermatogenesis within the Seminiferous Tubules

a. **View of the testes and the seminiferous tubules within.**

b. **Cross section of seminiferous tubules (4X).**

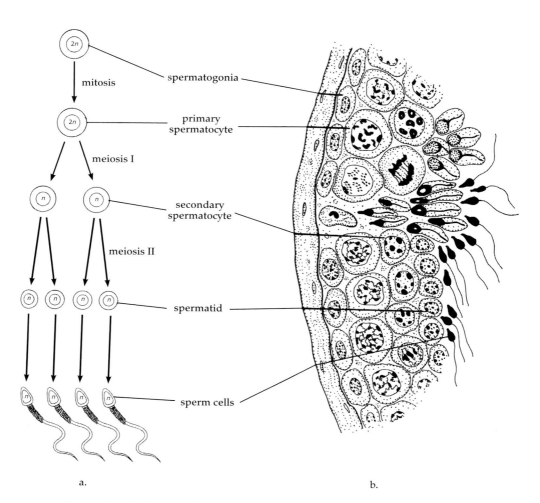

a. Stages of meiosis      b. Section through one seminiferous tubule

**FIGURE 14-6.** Meiosis within the Seminiferous Tubules

**Exercise 14 ■** *The Reproductive System*

5. The **spermatogonia divide by mitosis**, producing two daughter cells with each division. Some of these daughter cells remain as spermatogonia and others develop into **primary spermatocytes**. The primary spermatocytes begin **meiosis I**, which produces **two secondary spermatocytes**.

   Look for the **primary spermatocytes** in a layer just inside that of the spermatagonia (closer to the lumen). The primary spermatocytes are larger than the spermatagonia.

   The next layer contains **secondary spermatocytes**, which are undergoing meiosis II. Meiosis II produces **four** daughter cells called **spermatids**. Each is **haploid**, containing only **23 chromosomes** (half the number of a normal body cell).

   The spermatids differentiate into **sperm**. The tails of many can be seen projecting into the lumen of the tubule. Immature sperm continue their maturation in the epididymis, which lies just outside the testis.

## Comprehension Check

1. A house cat body cell has 38 chromosomes. If we were looking at the testes of a male house cat, how many chromosomes would be in a **spermatid**?

   How many chromosomes would be in the **spermatogonia**?

2. The process of spermatogenesis and maturation takes approximately nine weeks to complete. If two ejaculations occur within a short time period, will the second ejaculation contain sperm? **Explain** your answer.

3. Note that there are many more spermatids than secondary spermatocytes in each tubule. **Explain** why.

**Check your answers with your instructor before you continue.**

# ACTIVITY 3     DEVELOPMENT OF EGGS IN THE OVARIES

*Preparation*

In humans, egg production (**oogenesis**) occurs by **meiosis** in the follicles of the ovaries. At puberty, a woman has approximately 400,000 follicles. Most of the eggs inside these follicles, however, will never be ovulated, since only one egg is usually released per menstrual cycle.

In the following activity, you will observe the development of egg cells with the compound microscope.

1. Work with a group. Get **two compound microscopes, one dissecting microscope, and three cross section slides of the ovary: immature follicles, mature follicles, and corpus luteum**.

2. Position the two compound microscopes side by side on the laboratory table. Place the slides of **immature and mature follicles** on the stages of the microscopes.

   Observe the slide of the immature follicles with the scanning lens (**4X**). Close to the **outer edge** of the ovary, you will observe clusters of small follicles (see **Figure 14-7**). Inside each follicle is a **primary oocyte**, a developing egg. The primary oocyte surrounded by follicle cells is called a **primordial follicle**.

   Switch to the low power lens (**10X**) for a closer view of the primordial follicles.

   After puberty, under the influence of hormones, various groups of primordial follicles are activated and continue the stages of development listed below.

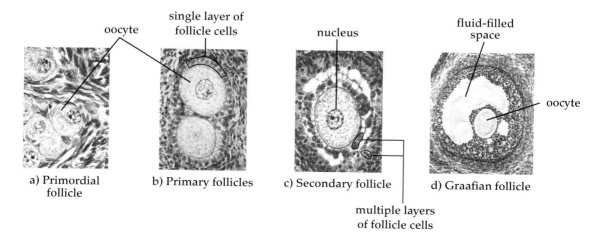

**FIGURE 14-7.** Stages of Follicle Development within the Ovary

**Exercise 14** ■ *The Reproductive System*

3. Near the clusters of primordial follicles, you will observe larger follicles, each with only a **single layer** of follicle cells surrounding the oocyte, which has also increased in size (see **Figure 14-7**). These are called **primary follicles**.

4. Locate a follicle with **several distinct layers of follicular cells** surrounding the oocyte. Switch to the low power lens (**10X**) for a closer view.

   These are the **secondary follicles**. As secondary follicles mature, spaces appear within the follicular layers. As you scan the slide, you will see secondary follicles in various stages of development.

   Cells of developing follicles produce the hormone **estrogen**. As the layers of follicle cells thicken, estrogen secretion increases.

   Move to the **second slide** (**mature follicles**).

5. Locate a **Graafian follicle** on the **mature follicle slide** (see **Figure 14-7**). The Graafian follicles are quite **large with a fluid-filled space** between the egg and the surrounding follicular cells (since this fluid is usually lost in slide preparation, the space will appear empty). The egg is often located on a small mound of follicle cells.

   **Switch back and forth between the two microscopes**, comparing the structure and appearance of the primary and secondary follicles with the Graafian follicles.

   Near the end of the development of the Graafian follicle, the primary oocyte inside the follicle completes meiosis I and begins meiosis II. Each daughter cell of meiosis is **haploid**, containing only **23 chromosomes** (half the number of a normal body cell).

   The Graafian follicle releases the egg and **ovulation** occurs. In humans, meiosis II only occurs after an egg has been fertilized.

   > **Note:**
   > Often the slides that are used to demonstrate oocyte development are obtained from animals that produce multiple young. Therefore, more follicles will be present than would be observed in human ovaries.

6. Place the slide of the **corpus luteum** on the stage of the **dissecting microscope**.

   The corpus luteum is an enormous circular or oval mass of cells. It may appear to form a bulge on the side of the ovary.

   The corpus luteum forms from the empty follicle remaining after ovulation. The cells of the corpus luteum produce the hormone **progesterone**, which prepares the uterus for pregnancy. The corpus luteum also continues to produce small amounts of estrogen, but progesterone is the primary hormone produced by this structure.

The corpus luteum continues to function for several months during pregnancy. If fertilization does not occur, however, the corpus luteum gradually degenerates, and progesterone production declines.

## ✓ Comprehension Check

1. A human body cell has 46 chromosomes. If we were looking at an **ovulated egg that has been fertilized**, how many chromosomes would there be from the original egg cell?

    How many total chromosomes would be present (from both the egg and sperm)?

2. Which would produce more estrogen: a primordial follicle or a secondary follicle? **Explain** your answer.

3. If ovulation occurs at approximately midpoint in the reproductive cycle, would you expect to find secondary follicles before or after ovulation?

    Would you expect to find a corpus luteum before or after ovulation?

    **Explain** your answers.

4. **Challenge question!** On a slide of the ovary, it is common to see large follicles with no eggs visible. Suggest an explanation for this observation.

# Exercise 14 ■ *The Reproductive System*

Self Test

1. Fill in the blanks with the name of the appropriate reproductive structure or its function.

| STRUCTURE | FUNCTION |
|---|---|
|  | production of sperm and sex hormones |
| follicle |  |
| seminiferous tubules |  |
|  | production of eggs and sex hormones |
|  | pouches of skin outside the abdomen that enclose the testes; provide a cooler environment for sperm development |
| epididymis |  |
|  | site of embryo implantation and fetal development |
|  | production of lubricating mucus and secretion of seminal fluid; provides a fluid transportation medium for the sperm |
| vas deferens |  |
|  | muscular tube that receives sperm; also serves as the birth canal |
| oviduct (fallopian tube) |  |
|  | muscular tube that carries urine from the urinary bladder to the outside; in males, also conveys semen during ejaculation |
| penis |  |

2. Correctly label the following structures on **Figure 14-8**: **urethra, vas deferens, testis, inguinal canal, umbilical arteries, pelvic bones,** and **urinary bladder**

**FIGURE 14-8.** Reproductive System of the Male Pig

**Exercise 14** ■ *The Reproductive System*

3. Correctly label the following structures on **Figure 14-9**: **urethra, vagina, uterine horns, umbilical arteries, ovary, pelvic bones, urinary bladder,** and **genital papilla**

**FIGURE 14-9.** Reproductive System of the Female Pig

4. Explain what happens to a ruptured follicle after ovulation.




5. If a woman has to have her uterus removed, her ovaries might be left in position because:




6. A fertility clinic in Germany has been advising its male clients to avoid wearing excessively tight jeans. Is there any physiological basis for this recommendation? **Explain** your answer.




7. How does the function of the urethra in males differ from its function in females? **Explain** your answer.




8. Explain why you would not expect to find a corpus luteum prior to puberty.

# EXERCISE 15

# Mitosis and Asexual Reproduction

## Objectives
After completing this exercise, you should be able to:
- name and describe the stages of mitosis
- identify the stages of plant and animal mitosis as viewed through the compound microscope
- discuss the relationship between the number of cells in each of the stages of mitosis and the length of the various stages
- explain the relationship between mitosis and the processes of regeneration and asexual reproduction
- apply your knowledge of asexual reproduction to cloning and other medical technologies.

## CONTENT FOCUS

Most of us are aware that the outer layer of our skin is subject to constant wear. To keep this protective layer intact, skin cells must be replaced throughout our lives.

Cells of the epidermis, for example, are replaced every 25–45 days. A similar situation occurs elsewhere in the body.

Growth is another situation in which additional body cells are required. A **zygote** (fertilized egg) begins life as a single cell that multiplies into many cells as the embryo develops. During childhood, cell multiplication provides the many cells needed for growth to an adult body size.

The type of **cell division** that makes growth and repair possible is called **mitosis**. All body cells produced by mitosis must contain **the same genetic information** as all other body cells—the information that makes you a unique individual.

# ACTIVITY 1  HOW MITOSIS WORKS

Preparation

A cell is genetically "programmed" to carry out its function in the body. These cellular instructions are included within structures referred to as **chromosomes**. A full set of these genetic instructions is necessary if a cell is to function normally. Therefore, when new body cells are produced by **mitosis**, each has a complete set of chromosomes.

The normal number of chromosomes in a cell is referred to as the **diploid number (abbreviated 2n)**. Each type of organism has a characteristic number of chromosomes. The diploid number of chromosomes in human body cells, for example, is 46. The diploid number in one species of pine tree is 24, in crayfish it is 200, and in fruit flies only 4.

Assume that the diploid number of chromosomes in the cell below equals 16.

If this cell divides, how many chromosomes must be present in each new cell?

**Enter this number** in each of the blank circles below.

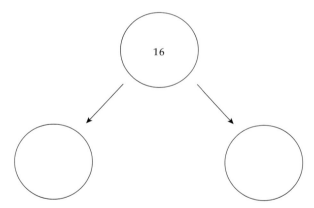

To make it easier to understand, the process of cell division is studied in five stages (called the **cell cycle**). At the completion of the cell cycle, two new cells are formed. Each cell will have a complete set of genetic instructions.

### Exercise 15 ■ *Mitosis and Asexual Reproduction*

To complete cell division successfully, **dividing cells must solve several problems:**

Each new cell needs a full set of **chromosomes**.

**Accomplished by:** Duplicating chromosomes (an exact copy is produced).

To avoid confusion, these duplicate chromosomes are referred to as **sister chromatids**.

A cell preparing to divide will contain two complete sets of sister chromatids (one set for delivery to each of the two new cells).

Sister chromatids must be attached together.

**Accomplished by:** A structure called a **centromere** fastens the duplicates together.

Attachment makes it easier to keep track of sister chromatids for sorting into two groups (one set for each new cell).

Pieces of chromosomes should not get broken off or lost.

**Accomplished by:** The long, threadlike chromosomes coil and fold into compact structures. Condensed chromosomes are thick enough to be visible with the compound microscope.

The pairs of sister chromatids must be pulled apart and delivered to each new cell.

**Accomplished by:** A structure called the **mitotic spindle** hooks on to each pair of sister chromatids. **Spindle fibers** pull the chromatids apart and deliver one chromatid from each pair to each new cell.

When the sorting and delivery process is complete, the two cells must be separated into two **daughter cells**, each with its own set of chromosomes.

**Accomplished by:** A process called **cytokinesis** separates the cytoplasm into two halves (**"cyto"** means *cell* and **"kinesis"** means *cutting*).

During cytokinesis in animal cells, the cell membrane pinches into a groove called the **cleavage furrow**.

## ACTIVITY 2      RECOGNIZING THE STAGES OF MITOSIS

Preparation

Use the **Decision Tree in Figure 15-1** to **identify the stages of the cell cycle** in **Figure 15-2**.

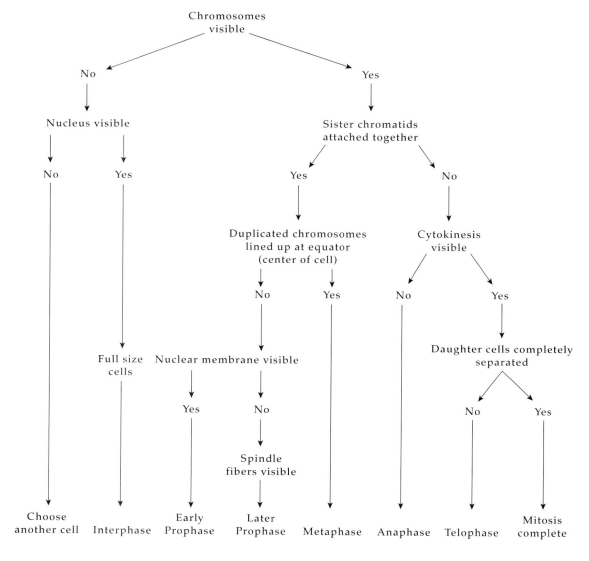

**FIGURE 15-1.** Mitosis Decision Tree

**Exercise 15** ■ *Mitosis and Asexual Reproduction* **277**

1. _____

2. _____

3. _____

4. _____

5. _____

6. _____

7. _____

**FIGURE 15-2.** Stages of the Cell Cycle to Identify

## Comprehension Check

1. Put the following stages of the cell cycle in the correct order (1 through 5).

    _____ Metaphase        _____ Telophase

    _____ Interphase       _____ Prophase

    _____ Anaphase

Use the following vocabulary words to fill in the blanks in the questions below. Each answer can be used **only once**.

**centromere**         **sister chromatids**         **cleavage furrow**
**cytokinesis**        **chromosomes**               **mitotic spindle**
**equator**

2. During metaphase, chromosomes line up along the _____ of the cell.

3. The two _____ are held together with a _____.

4. _____ refers to the division of the cytoplasm into two daughter cells.

5. Genetic information in cells is contained in structures called _____.

6. During anaphase, part of the _____ can be seen attached to the centromere of each sister chromatid.

7. The constricted appearance of the cell membrane prior to the formation of two daughter cells is called the _____.

**Check your answers with your instructor before you continue.**

Exercise 15 ■ *Mitosis and Asexual Reproduction* 279

# ACTIVITY 3                                    MITOSIS—THE REAL THING!

*Preparation*

The stages of the cell cycle in living cells are often not as clear as in a set of diagrams. Sharpen your observation skills and practice recognizing the stages of mitosis at the same time.

1. For this part of the activity, **work on your own**. Get the following supplies: one **slide of mitosis in whitefish embryos** and a **compound microscope**.

2. View the slide with the **high-power lens (40×)**.

3. Your instructor will assign you **one stage of the cell cycle** to locate on your slide.

   When you locate a **clear** example of that stage, indicate the cell with the **pointer** in the ocular lens.

 **Check your answers with your instructor before you continue.**

4. Draw the stage **as it appears under the microscope**, in the appropriate location on **Figure 15-3**.

   In your drawing, label the following (if present): **chromosomes, mitotic spindle, equator, cleavage furrow, cell membrane,** and **nuclear membrane**.

5. After completing your drawing, **observe** the other **stages of the cell cycle** as identified by other students.

   **Create your own drawings** of these stages and add them to the appropriate locations on **Figure 15-3**.

   In your drawings, label the following (if present): **chromosomes, mitotic spindle, equator, cleavage furrow, cell membrane,** and **nuclear membrane**.

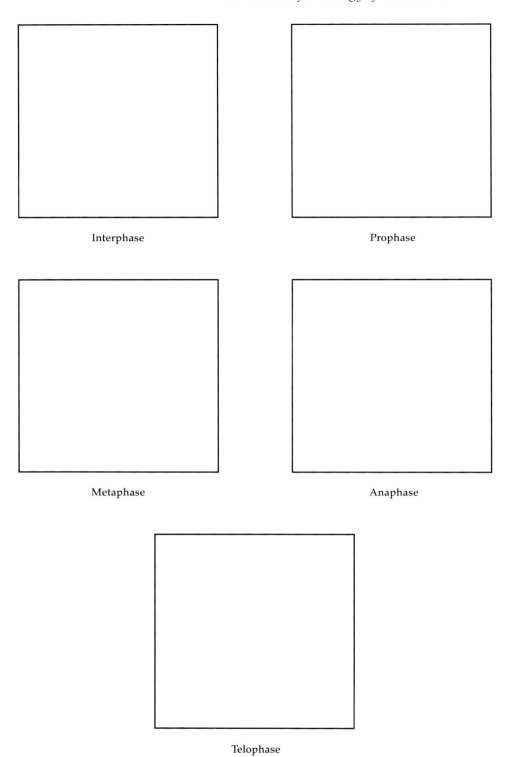

**FIGURE 15-3.** The Cell Cycle as Seen through the Microscope

# Exercise 15 ■ *Mitosis and Asexual Reproduction*

## ACTIVITY 4 — ESTIMATE THE DURATION OF THE CELL CYCLE

*Preparation*

When you look at a prepared slide, you are looking at a "moment frozen in time." The preserving chemicals stopped the cells in the middle of their daily activities. From this slide, you can see how many cells were undergoing cell division at that moment. You can even see how many cells were in each of the stages of mitosis. You can use this information to estimate the length of each stage of the cycle compared to the other stages.

1. For this activity, **work individually**. Get the following supplies: one **slide of onion root mitosis** and a **compound microscope**.

   View the slide with the **high-power lens**. Position your view of the onion root tip to a section where many cells are undergoing mitosis (see **Figure 15-4** for example).

   Your instructor will give you additional instructions on how to position the slide for optimal viewing.

2. Using the stages of onion root tip mitosis in **Figure 15-4** as a guide, **count** the number of cells **in your field of view** that are in each stage of the cell cycle.

   Record your results in **Table 15-1**.

### TABLE 15-1
### RESULTS OF CELL CYCLE OBSERVATIONS

| NUMBER OF CELLS IN STAGE | INTERPHASE | PROPHASE | METAPHASE | ANAPHASE | TELOPHASE |
|---|---|---|---|---|---|
| View 1 | | | | | |
| View 2 | | | | | |
| View 3 | | | | | |
| Total for Views 1–3 | | | | | |
| Percentage of Cell Cycle | % | % | % | % | % |

| Interphase | Prophase | Metaphase | Anaphase | Telophase |

**FIGURE 15-4.** Stages of Mitosis in the Onion Root Tip

3. Repeat the **instructions in Step 2 twice more** (so you will examine a total of **three different onion fields of view**), and record the results for each in **Table 15-1**.

4. **Total** the number of cells counted for each view and record the totals in **Table 15-1**.

5. **Calculate the percentage of cells observed in each stage of the cell cycle.**

STEP 1:

Add the total number of cells counted in all five stages.

Total Cells Counted = Interphase + Prophase + Metaphase + Anaphase

+ Telophase

Total Cells Counted = \_\_\_\_

STEP 2:

Divide the number of cells in Interphase by the total from Step 1 and multiply by 100.

$$\frac{\text{Number in Interphase}}{\text{Total Cells Counted}} \times 100 = \underline{\phantom{000}}\%$$

STEP 3:

Repeat Step 2 for the remaining four stages of the cell cycle. Record all percentages in **Table 15-1**.

**Exercise 15** ■ *Mitosis and Asexual Reproduction*

6. Using the information in **Table 15-1**, form a **hypothesis** about the **length** of each stage of the cell cycle.

   Which do you think is the longest stage? _____

   Which do you think is the shortest stage? _____

   Do you think any stages are about the same length? If so, list them below.




   **Explain how you formed your hypothesis.**




7. See **Table 15-2 (after the Self Test)** for the results of some experiments on the length of the onion cell cycle. Do the experimental results support your hypothesis? _____. **Explain your answer**.


| TABLE 15-2 RESULTS OF EXPERIMENTS ON THE LENGTH OF STAGES IN THE ONION CELL CYCLE | | | | | |
|---|---|---|---|---|---|
| | INTERPHASE | PROPHASE | METAPHASE | ANAPHASE | TELOPHASE |
| Length of Stage | 17 hrs | 88 min | 4 min | 3 min | 6 min |
| Percentage of Cell Cycle | 91% | 7.9% | 0.4% | 0.3% | 0.5% |

## ACTIVITY 5 — REGENERATION

Preparation

Some animals have the ability to replace (regrow) lost or damaged body parts. This process is known as **regeneration** and is a form of asexual reproduction. You may be familiar with some animals that have this ability. Starfish (also known as **sea stars**) can regenerate body parts as long as a small part of the central disk is present. Starfish can grow a new arm and the severed arm can grow a whole new body! If cut in two, each half of the starfish can develop into a whole individual.

During the microscope exercise, you looked at planaria, freshwater flatworms. Planaria also have the ability to regenerate lost body parts. Any piece of the body except the tail can regenerate a complete new adult. You will cut **planaria** to demonstrate regeneration and regrowth.

1. Work in groups. Get the following supplies: **one clean razor blade in a plastic holder, a glass petri dish, a pipette,** and **crushed ice**.

2. You will also need a bottle of **pond water, labeling tape**, and a **dissecting microscope**.

   Set up your dissecting microscope, but keep the light **off**. On the stage of the microscope, place the **top half** of the petri dish.

   Fill the dish **to the rim** with crushed ice.

3. Place **one large planaria** into the **bottom half** of your petri dish with a small amount of **pond water**.

   Set the dish with the planaria **onto the crushed ice**. This will anesthetize the worm for the "operation."

   Let the worm relax on the ice for about **five minutes**.

4. While you are waiting, decide how you will cut the planaria. You can choose one of the two cutting patterns in **Figure 15-5**, or develop your own pattern.

5. Make **perpendicular cuts** with the razor blade through the worm. Do not tilt the razor blade when cutting. The cuts must be sharp and clean.

6. After the operation, add some clean **pond water** to the petri dish containing the worm.

   Discard the ice. Wash and dry the top half of the petri dish. Using labeling tape, mark the lid with **your laboratory section and your name**.

**Exercise 15** ■ *Mitosis and Asexual Reproduction*

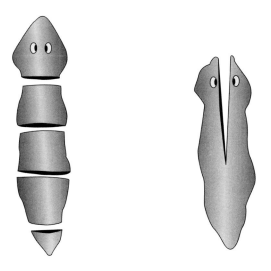

**FIGURE 15-5.** Planaria Cutting Patterns

Return the petri dish containing the worm to your instructor to be stored in a cool, quiet location.

7. After **two weeks**, you can examine your worm under the dissecting microscope to see the results of your experiment.

8. Challenge Question! If you cut a planaria into five pieces and each piece developed into a separate individual, would you be able to tell them apart? **Explain your answer.**

## ACTIVITY 6 — ASEXUAL REPRODUCTION: GROWING A PLANT FROM A CUTTING

Preparation

As you have seen, mitosis provides genetically identical cells for growth, replacement, and repair of body tissues. Mitosis also provides a mechanism for **reproducing whole organisms asexually**. In asexual reproduction, offspring are produced, which are **genetically identical** to the single parent and also to one another.

Plants do not have to come from seeds. You can remove cuttings from your plants. The cuttings will develop roots and grow into new plants. This form of asexual reproduction is an example of **plant propagation**.

1. Get **a paper cup and a scalpel**.

   You will also find **plants for propagation, rooting hormone,** and **potting soil**.

2. Fill the cup with **potting soil** up to **one inch** from the top of the container.

   **Moisten** the soil with water until it is damp, but not soggy.

3. **Cut a small piece of stem** with several leaves attached.

   Dip the **cut end** of the stem into the **rooting hormone** and **plant** it in the cup.

4. Keep your baby plant **moist** and exposed to **bright, indirect light**. Roots should form within several weeks.

## ✓ Comprehension Check

1. A botanist is trying to save a rare plant from extinction. Suggest a method she can use to increase the population.

2. The **type of cell division** that produced root growth in my cutting is _____.

3. Dolly, an internationally famous sheep, was conceived by an unusual process. Dolly was **cloned**—a type of **asexual reproduction**. What percent of Dolly's characteristics would be similar to her mother's? _____%. **Explain your answer.**

4. How would a colony of mice produced by **cloning be helpful** for testing the effectiveness of a new diet pill?

**Check your answers with your instructor before you continue.**

**Exercise 15** ■ *Mitosis and Asexual Reproduction*

Self Test

Fill in the blanks with the most appropriate answer. Answers can be used **only once**.

a. Diploid
b. Sister chromatid
c. Centromere
d. Spindle fibers
e. Nuclear membrane
f. Cell membrane
g. Cytokinesis
h. Cleavage furrow
i. Equator
j. Zygote
k. Daughter cell
l. Chromosome

1. _____ Breaks down at the end of prophase, releasing the chromosomes.

2. _____ Chromosomes are arranged along this imaginary line during metaphase.

3. _____ Structure that holds two duplicate chromosomes together.

4. _____ Two of these are found at the **end** of mitosis.

5. _____ Normal number of chromosomes in a cell.

6. _____ Responsible for chromosome movement during mitosis.

7. _____ A constriction of the cell membrane that occurs in telophase.

8. _____ Duplicated chromosome has two of these.

9. _____ Process that separates the cytoplasm into two halves.

10. _____ Genetic material found in the cell nucleus.

Fill in the blanks with **true** or **false**.

11. _____ If a cell undergoing **mitosis** has a chromosome number of six, the daughter cells will have a chromosome number of three.

12. _____ In asexual reproduction, the offspring are identical to the parent.

13. _____ If cells are multiplying, as they do when grown in tissue cultures, mitosis is taking place.

14. _____ Growth during childhood is a good example of cell division.

15. Complete mitosis for the following parent cell. Draw each stage of mitosis **in order**, and show the number of chromosomes in each stage (including the two daughter cells). The beginning cell has a diploid number of four chromosomes.

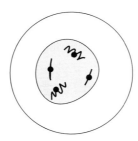

16. You are vacationing in Maine. One morning, while walking near the shore, you hear a group of fisherman complaining that there are too many starfish in the ocean. They plan to solve this problem by netting the starfish, cutting them in half to kill them, and throwing them back into the ocean. Is their plan likely to be effective? **Explain your answer.**

17. *Challenge Question!* If you wanted to asexually propagate a human, you could use

    a. a skin cell
    b. a sperm cell
    c. a red blood cell
    d. any of the above would be suitable
    e. none of the above would work out

    **Explain your answer.**

# EXERCISE 16

# Connecting Meiosis and Genetics

## Objectives
After completing this exercise, you should be able to:
- name and describe the stages of meiosis
- correctly use and understand the terminology associated with cell division and genetics
- demonstrate an understanding of the changes in chromosome number that occur during meiosis and fertilization
- compare meiosis I with meiosis II in terms of the position of the chromosomes in each stage, changes in chromosome number, and number of daughter cells produced
- explain the process and importance of crossing over between homologous chromosomes
- draw and complete Punnett squares and use them to determine genetic probabilities in monohybrid crosses
- solve genetics problems involving other modes of inheritance, including incomplete dominance, codominance, and multiple allele traits.

## CONTENT FOCUS

Your body cells contain **46 chromosomes** within the nucleus. As you probably k[now], **half** of this genetic information (23 chromosomes) is inherited from your moth[er and] **half** (the other 23 chromosomes) from your father.

The genetic information carried by the **gametes** (the sperm and egg), wh[en incor]porated into a fertilized egg, will determine all the physical and physiologi[cal traits of] the offspring.

There is a special type of cell division that changes the chromosome [number from] the normal **diploid** number (46) to the **haploid** number (23) found in s[perm and egg.] This special type of cell division is called **meiosis**.

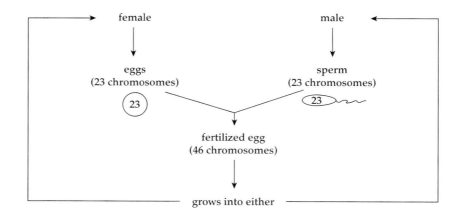

The stages of meiosis, in most respects, are similar to those of mitosis. The important differences include the following:

- the chromosome number is reduced from diploid to haploid to form gametes
- an exchange of genetic material takes place in a process known as **crossing over**; gametes produced by meiosis are **not** genetically alike
- meiosis involves two cell divisions (meiosis I and meiosis II); one occurs immediately after the other.

The genes on a chromosome may exist in **more than one form**, called **alleles**. Individuals inherit **two alleles for each trait**, one received from the mother in the egg and the other from the father in the sperm.

two inherited letters represent **different** forms of the gene (different individual is called **heterozygous for that trait**. The allele that is heterozygous individuals is referred to as **dominant** and is represented (**B**).

essed in heterozygous individuals is called **recessive**. Recessive ith lowercase letters (**b**).

dentical alleles for a trait (**BB or bb**), they are said to be

it is an individual's **genotype** (such as BB or Bb).

rait is called the **phenotype** (such as brown

the process of meiosis to **form** male and ormation carried by these gametes, and use face.

**Exercise 16** ■ *Connecting Meiosis and Genetics*

5. **Repeat step 5** to connect the alleles of the **paternal chromosome** using the **red centromere**.

    Together these **two chromosomes (maternal and paternal)** are referred to as **homologous chromosomes**. Homologous chromosomes carry alleles for the **same traits** (face shape, eye size, etc.), although the genetic information is **not identical**.

6. Fill in **Table 16-2** with the genotypes and phenotypes of the traits on the **chromosomes** you have just constructed.

    Solid beads = dominant alleles. Striped beads = recessive alleles.

| TABLE 16-2 WHAT ARE YOUR TRAITS? | | |
|---|---|---|
| TRAIT | YOUR GENOTYPE | YOUR PHENOTYPE |
| face shape | FF | round |
| hair | Hh | wavy |
| eye size | Ee | medium |
| eye distance | Dd | medium |
| eyebrow shape | Bb | thick |
| eyelash length | Ll | long |
| nose size | nn | small |
| lips | Gg | thick |
| earlobes | Rr | free |
| cleft in chin | tt | absent |
| freckles | Qq | freckles |

## ACTIVITY 2 — HOMOLOGOUS CHROMOSOMES SEPARATE IN MEIOSIS TO FORM GAMETES

1. Draw a series of circles that are the same as those shown in **Figure 16-2**.

   **Make the circles big enough for your chromosomes to fit comfortably.**

2. Place your pair of chromosomes in the first circle. Your cell is now in **interphase**.

   What is the diploid number of chromosomes in this cell? _____

   How many chromosomes came from your mother? _____ How many from your father? _____

   How many **traits** are represented on your chromosomes? _____

   How many **alleles** are represented on your chromosomes? _____

3. An important event that occurs during **interphase** involves the replication of chromosomes. Use the spare beads in your genome bag to **replicate** your two chromosomes.

   Two identical (replicated) DNA strands are called **sister chromatids**.

   Attach the new **sister chromatids** for your **maternal chromosome** together at the centromeres. Do the same for your replicated **paternal chromosome**.

4. Move your duplicated chromosomes to the next circle, marked **prophase** through **telophase of meiosis I**.

   You will complete **prophase through telophase of meiosis I** in the **same circle**.

5. As shown in **Figure 16-1**, during **prophase I**, the **two homologous chromosomes** find each other and pair up in a process called **synapsis.**

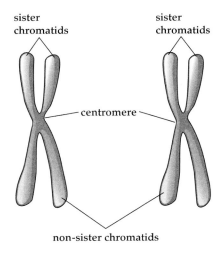

**FIGURE 16-1.** Duplicated Homologous Pair (Tetrad)

6. During synapsis, the **non-sister chromatids** exchange genetic information. This process is called **crossing over.**

   **Simulate crossing over** between **two non-sister chromatids** in your homologous pair (exchange of alleles between **one maternal and one paternal** chromosome).

   Exchange alleles for the **last five traits** (nose size through freckles).

   *Note:*
   **Exchange alleles ONLY for two NON-SISTER chromatids!**

7. Simulate **metaphase I** by placing the chromosomes **in the correct position** relative to the **equator.**

   To simulate **anaphase I**, separate the two homologous chromosomes by moving them to the opposite poles of the cell.

   To simulate the division of the cytoplasm in **telophase I**, draw a dotted line that represents the separated **daughter cells.**

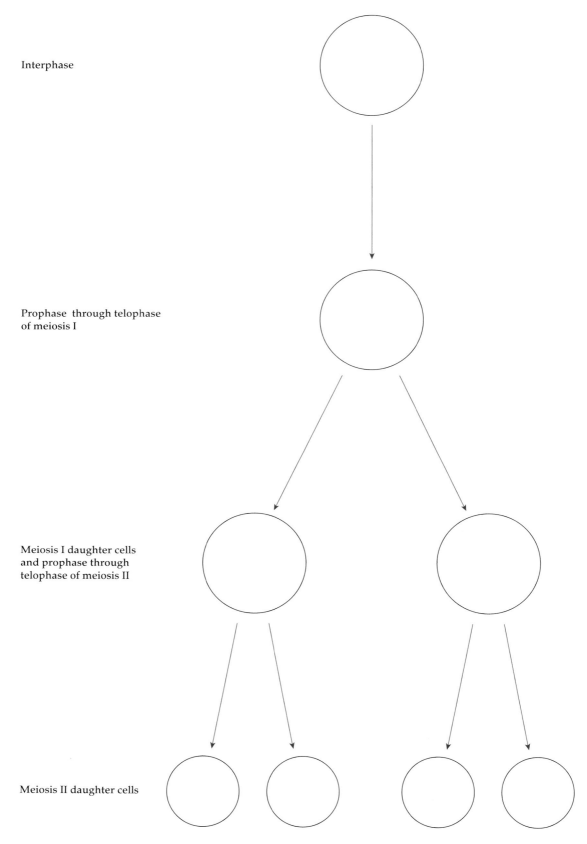

**FIGURE 16-2.** Pattern for Meiosis Simulation

**Exercise 16 ■** *Connecting Meiosis and Genetics*

8. Move your chromosomes to the next two circles that are labeled **Meiosis I daughter cells.**

   ✓ Comprehension Check

   1. What is the number of chromosomes in each daughter cell? _____

   2. Does this number represent the **diploid** or the **haploid** chromosome number?
      _____

   3. What happened to the **maternal and paternal** chromosomes from the original parent cell?

   4. Are the two daughter cells genetically identical? _____ **Explain your answer.**

9. Meiosis continues with a second cellular division.

   Simulate meiosis II without moving your chromosomes to another set of circles. Note that the circles you are now using have two labels. In **addition** to Meiosis I daughter cells, the circles are also labeled **prophase through telophase of meiosis II.**

   Nothing unusual occurs to the chromosomes in prophase II.

   To simulate **metaphase II**, place the chromosomes **in the correct position** relative to the **equator**.

    To simulate **anaphase II**, separate the two **sister chromatids** by moving them to the opposite poles of the cell.

   Move your chromosomes to the four circles that are labeled **Meiosis II daughter cells.** Each of these daughter cells has the potential to develop into a sperm or egg.

 **Check the movement of your chromosomes through meiosis I and II with your instructor before you continue.**

## Comprehension Check

1. The original parent cell in your meiosis simulation had two alleles for each trait.

    How many alleles for each trait are **now** in each daughter cell? _____

2. Are your daughter cells diploid or haploid? _____

3. Are the four daughter cells **genetically identical?** _____ **Explain your answer.**

4. Where in the body does meiosis occur in **males**? _____ Where in **females**? _____

## ACTIVITY 3 — FERTILIZATION: NATURE'S EQUIVALENT TO ROLLING THE DICE

1. As you know, **only one sperm and one egg** can participate in fertilization.

    Any of the four sperm can potentially fertilize an egg, but the development of female gametes is slightly different. **Only one** of the four daughter cells will develop into an egg cell. The other three are not functional and are referred to as **polar bodies**.

    You will simulate fertilization of your gametes with a roll of the dice. Get a **single six-sided die**.

    Are **your** gametes potential sperm or eggs? _____

2. Choose the gamete that will participate in fertilization as follows:

    a. **Number your daughter cells** 1 through 4.

    b. **Roll the die.** If you get a number between 1 and 4, that is your lucky gamete. If you get 5 or 6, roll again.

3. Link up with a group that has produced a gamete of the opposite sex. Take the chromosomes from both gametes and **place them in a circle** labeled "**Fertilized Egg.**"

Exercise 16 ■ *Connecting Meiosis and Genetics* **299**

# ACTIVITY 4                         WHAT IS YOUR BABY'S GENOTYPE?

1. On the two chromosomes in **Figure 16-3**, list your baby's **alleles in their correct order**.

   Solid beads = dominant alleles. Striped beads = recessive alleles.

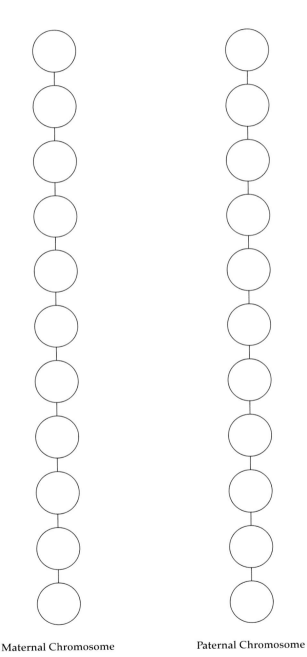

Maternal Chromosome          Paternal Chromosome

**FIGURE 16-3.** Maternal and Paternal Chromosomes of the Fertilzed Egg

## ACTIVITY 5 — GENOTYPE DETERMINES PHENOTYPE

1. Now that you know your baby's **genotype** for each trait, turn back to **Table 16-1**. In **Table 16-1**, **highlight or circle** your baby's **genotype and phenotype** for each trait.

   Now that you know your baby's **phenotype** for each trait, you are ready to see what your baby will look like!

2. Choose the appropriate **face shape** outline to begin your baby (**Figures 16-4** and **16-5**).

   **Cut out** the appropriate **eyes, ears, and other facial features** and use **tape** to attach them to your baby's face (**Figures 16-6 through 16-13**). If your baby has freckles, **draw them** in.

   **Add** the appropriate **eyelashes** to your baby's face.

   Are you pleased with your results? If not, remember that each time a sperm fertilizes an egg you get a different combination of alleles. You may have better luck next time.

3. **Roll the die again** to determine if your baby is a boy or a girl. If you roll **1 through 3**, it is a **girl**. If you roll **4 through 6**, it is a **boy**.

**Exercise 16** ■ *Connecting Meiosis and Genetics* 301

**FIGURE 16-4.** Round Face

**FIGURE 16-5.** Triangular Face

**FIGURE 16-6.** Curly Hair

**FIGURE 16-7.** Wavy Hair

# Exercise 16 ■ Connecting Meiosis and Genetics

**FIGURE 16-8.** Straight Hair

Large Eyes

Medium Eyes

Small Eyes

**FIGURE 16-9.** Eye Types

**Exercise 16** ■ *Connecting Meiosis and Genetics* **307**

Thick Eyebrows                    Thin Eyebrows

**FIGURE 16-10.** Eyebrow Shape

Thick Lips                    Thin Lips

**FIGURE 16-11.** Lip Types

Large nose

Medium nose

Small nose

**FIGURE 16-12.** Nose Size

# Exercise 16 ■ Connecting Meiosis and Genetics

**FIGURE 16-13.** EarLobes

# Exercise 16 ■ Connecting Meiosis and Genetics

# ACTIVITY 6                                         PASSING ON TRAITS

Preparation

Twenty-four years have passed. Your baby is now an adult and has married someone who is **heterozygous for freckles**. What is the probability that they will have a child with freckles?

The first step in calculating the probability of freckles is to **determine the alleles carried by the sperm and eggs** of this couple.

As you know, alleles separate in meiosis, so that each sperm and each egg has only **one allele from each homologous pair**.

Your child's genotype for the freckle trait _____

1. A **Punnett Square** is a convenient way to visualize the different combinations of alleles that might occur during fertilization.

2. The spouse's alleles have been entered on the top row of the Punnett Square. Enter the alleles for **your child** along the left side of the square.

3. Using the Punnett Square as a guide, fill in the boxes by carrying the alleles down from the top and across (from left to right).

Each box in the Punnett Square represents a **25% probability** of having a baby with that specific genotype.

What is the probability that this couple will have a child with freckles? _____

4. **Assume that your baby has wavy hair** (inherited through **incomplete dominance**) and marries a person who **also has wavy hair**. Incomplete dominance is a special type of inheritance that occurs when an allele exerts only partial dominance over another allele. This results in an intermediate (blending) phenotype in heterozygous individuals. Complete the Punnett Square below to answer the following questions:

   What is the probability of a child with **straight** hair? _____

   What is the probability of a child with **curly** hair? _____

|   |   |
|---|---|
|   |   |
|   |   |

## ACTIVITY 7 — OTHER MODES OF INHERITANCE

Preparation

Now that you are beginning to be an expert in solving genetics problems, let's try something a little more challenging. Up to this point, we have dealt with traits that have **only two alleles.**

Human blood types (A, B, and O) are inherited by **multiple alleles**. There are **three possible alleles** for blood type, and **two** of those alleles are **codominant**.

Codominance is a special type of inheritance in which two alleles are equally dominant. Both alleles are expressed independently of each other, resulting in a heterozygous individual that shows both homozgous phenotypes. An example would be red blood cells with both A and B proteins in their cell membranes.

**Multiple alleles** means that there are **more than two possible alleles** for that trait, although each person will inherit **only two** of those alleles (one from the mother and one from the father).

In human blood types, the alleles $I^A$ and $I^B$ are codominant. Both $I^A$ and $I^B$ are **dominant** over the **recessive allele i**.

| Genotype | Phenotype |
|---|---|
| $I^A I^A$ or $I^A i$ | type A blood |
| $I^B I^B$ or $I^B i$ | type B blood |
| $I^A I^B$ | type AB blood |
| ii | type O blood |

EXERCISE 17

# Useful Applications of Genetics

## Objectives
After completing this exercise, you should be able to:
- demonstrate an understanding of the limitations of sample size in scientific data analysis
- determine genotypes using pedigree charts
- explain why more males than females express X-linked traits
- solve genetics problems involving dominant–recessive inheritance, X-linked traits, and codominance
- explain the evolutionary relationship between sickle-cell disease and malaria
- apply your knowledge of genetics to real-life situations.

## CONTENT FOCUS

Not all alleles produce visible traits like skin color or height. Most alleles control **physiological** traits, such as production of digestive enzymes, hormones, and antibodies. Alleles are responsible for invisible traits such as blood type, ability to carry out metabolic pathways (such as producing proteins or storing blood sugar), color vision, and many others. One example of a physiological trait controlled by a **single gene** is the ability to taste a harmless chemical, **PTC** (phenylthiocarbamide).

## ACTIVITY 1                                             PTC TASTING

1. Get the following supplies: one piece of **control** taste paper and one **containing PTC**.

2. **Taste the control paper** first (to establish the taste of the paper itself). Then **taste the PTC paper.**

If you are a taster, you will detect a very unpleasant, bitter taste. If there is little difference between the PTC and the control papers, you are not a taster.

Are you a taster or a nontaster? _____

3. Record your taste-test results **on the master chart at the front of the room.**

4. In the general population, approximately 75% of people can taste PTC. The remaining 25% are not able to taste this chemical.

   Were your class results close to the **expected percentage?** _____

   **If not**, suggest a possible reason why your class results differed from the expected outcome:

5. The ability to taste PTC comes from a **dominant allele (T).** Using this information, fill in the appropriate genotypes for tasters and nontasters.

   Tasters _____    _____

   Nontasters _____

6. The inheritance pattern of traits like PTC tasting can be diagrammed in a chart called a **pedigree.** A pedigree illustrates the marriages for several generations within a family and the children produced.

   - Females are shown with circles and males are shown with squares.
   - A **black square** or a circle shows the **presence of the condition** being studied. A **white square or a circle** means the **condition is absent** in that person.
   - A marriage or mating is shown by a line connecting the parents.
   - Children from a mating are shown by a vertical line between the parents.
   - All individuals from the same generation are shown along the same horizontal line.

**Exercise 17** ▪ *Useful Applications of Genetics*

7. After examining the **pedigree key**, determine the genotypes of **all the people** in this family. In a few cases, there may not be enough information to determine a person's second allele. In this situation, **enter a question mark (?) in place of the second letter.**

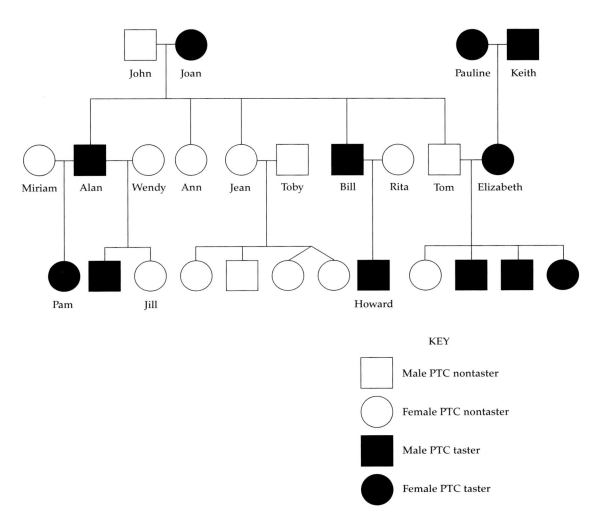

**FIGURE 17-1.** PTC Pedigree

8. Referring to the pedigree in **Figure 17-1**, if Jill marries Howard, what is the **probability** that their children will be

   tasters? _____ nontasters? _____

   **Show your work!**

9. If Pam marries Howard, what is the probability that their children will be tasters? _____ nontasters? _____

**Show your work!**

Check your answers with your instructor before you continue.

## ACTIVITY 2     TASTING OF OTHER CHEMICALS

*Preparation*

Since the ability to taste PTC is genetically based, it is likely that the ability to taste other substances can also be genetically determined. Our preferences for certain foods and dislike for others is determined, in part, by the types of chemicals we taste.

An example is the **artificial sweetener, Nutrasweet**, used in diet foods and beverages. Nutrasweet may taste **sweet or bitter** (depending on the individual). Another example is **sodium benzoate**, a **food preservative**. You may see it listed among the ingredients on the labels of many packaged foods. When sodium benzoate is tasted alone, tasters may describe it as **salty, sweet, sour, or bitter**.

**Do you think there might be a relationship between tasting PTC and tasting sodium benzoate? Write a hypothesis** about the relationship between PTC tasting and tasting sodium benzoate.

| CHEMICAL TASTING HYPOTHESIS |
|---|
|  |

Check your hypothesis with your instructor before you continue.

**Exercise 17** ■ *Useful Applications of Genetics*  321

# ACTIVITY 3 — TESTING YOUR HYPOTHESIS

> **Note:**
> There are not enough people in your laboratory group to give you an adequate test sample. People from other groups will be happy to volunteer as test subjects.

1. Use **Table 17-1** to record your data.

**TABLE 17-1 CHEMICAL TASTING DATA**

| Subject | | |
|---|---|---|
| 1 | | |
| 2 | | |
| 3 | | |
| 4 | | |
| 5 | | |
| 6 | | |
| 7 | | |
| 8 | | |
| 9 | | |
| 10 | | |
| 11 | | |
| 12 | | |

2. In a few sentences, **summarize the results** of your experiment.

3. Was your hypothesis **supported**? _____

4. **Write a conclusion based on your hypothesis and the collected data. Support your conclusion** by mentioning facts collected during your experiment.

✓ Comprehension Check

1. After performing this experiment, do you think that the ability to taste certain chemicals can influence your food preferences? **Explain your answer**.

2. **Barium sulfate** is tasteless to some people but bitter to others. Two sisters are having their gastrointestinal tracts x-rayed. They are both asked to drink a barium sulfate "milk shake" before the procedure. One sister drinks the "milk shake" without complaint. The other sister complains that it tastes terrible. Using the information gained in today's lab activities, **suggest an explanation for this difference**.

**Exercise 17** ■ *Useful Applications of Genetics*

# ACTIVITY 4 — X-LINKED TRAITS

*Preparation*

In humans, **the X chromosome is large in comparison to the Y chromosome**. The X chromosome carries information for many traits that are not related to the gender of the individual. Alleles carried only by the X chromosome are said to be **X-linked** (or sometimes, sex-linked).

Some of the alleles on the tiny **Y chromosome** appear to have no counterparts on the X. These **Y-linked alleles** code for traits that are found **only in males**.

Among the X-linked traits are a number of recessive genetic disorders. One of these is **hemophilia**, the inability to produce proteins necessary for blood clotting. Hemophiliacs may bleed to death from relatively minor cuts or bruises. Historical records dating back thousands of years mention the inheritance pattern of hemophilia. Among the ancient Hebrews, sons born to women with a family history of hemophilia were excused from circumcision.

Hemophilia was common during the 1800s in the royal families of Europe, whose members often intermarried. Queen Victoria of England was a **carrier** of the trait. She had one X chromosome with the allele for normal blood clotting ($X^H$) and the other with the defective allele ($X^h$). Because she did have **one normal dominant allele**, her blood clotted normally.

Her husband, Prince Albert, was completely normal for this trait. He had one normal allele on the X chromosome ($X^H$) and a Y chromosome with **no allele related to blood clotting** ($Y^o$).

Eighteen of Queen Victoria's 69 descendants were carrier females or hemophiliac males. Crown Prince Alexis of Russia was one of these hemophiliac descendants. His affliction indirectly contributed to the overthrow of the monarchy in Russia.

1. Complete this Punnett Square for the marriage of Victoria and Albert.

|  | Queen Victoria | |
|---|---|---|
|  | $X^H$ | $X^h$ |
| Prince Albert  $X^H$ |  |  |
| $Y^o$ |  |  |

2. What is the probability that Victoria and Albert could have

   a hemophiliac son           _____
   a hemophiliac daughter      _____
   a normal son                _____
   a carrier daughter          _____
   a daughter with normal blood clotting _____

3. Shown below is a partial pedigree of hemophilia in the descendants of Victoria and Albert. **Affected individuals have darkened circles or squares.** For **each individual in the pedigree**, fill in his or her probable genotype.

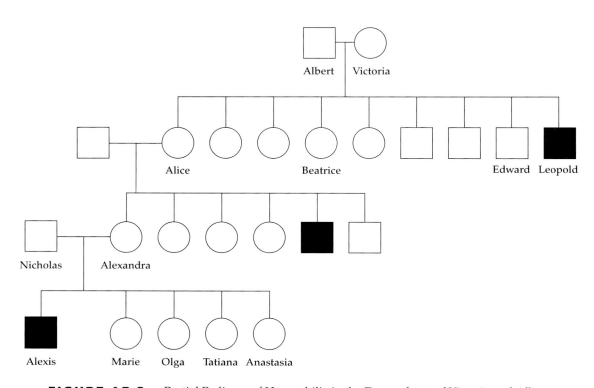

**FIGURE 17-2.** Partial Pedigree of Hemophilia in the Descendants of Victoria and Albert

Exercise 17 ■ *Useful Applications of Genetics* 325

### ✓ Comprehension Check

1. *Challenge Question!* If Victoria and Albert had a hemophiliac daughter, why would it be **unlikely for her to survive puberty** without medical intervention?

## ACTIVITY 5    CODOMINANCE AND SICKLE-CELL ANEMIA

Preparation

**Codominance** is a special type of inheritance in which two alleles are equally dominant, and, therefore, heterozygous individuals show **both homozygous traits**.

**Sickle-cell anemia** is an example of a human trait that is inherited through codominance. This condition is quite common in countries that have a high incidence of malaria. Sickle-cell anemia is a blood disorder affecting oxygen transport by hemoglobin. The sickled cells block capillaries, depriving the tissues of needed oxygen.

Individuals with sickle-cell anemia frequently die at an early age. **Heterozygous** individuals have **sickle-cell trait.** These carriers are usually healthy, but experience some problems with intense exercise or under low-oxygen conditions. In the United States, sickle-cell trait affects 1 out of 12 African Americans.

The allele for normal hemoglobin is $Hb^A$ and the sickle-cell allele is $Hb^S$. The following is a summary of the possible sickle-cell genotypes and phenotypes.

$Hb^A Hb^A$      completely normal

$Hb^A Hb^S$      sickle-cell trait (this person has a combination of normal hemoglobin and the abnormal, sickled form of hemoglobin)

$Hb^S Hb^S$      sickle-cell anemia (all abnormal hemoglobin)

1. What are the possibilities for children if **both parents are heterozygous** for sickle-cell trait?

|  |  |
|--|--|
|  |  |
|  |  |

2. What is the probability of a couple having a child with sickle-cell **trait** if **one** parent is **normal** and the **other** has **sickle-cell trait?**

|  |  |
|--|--|
|  |  |
|  |  |

# ACTIVITY 6     A FAMILY HISTORY OF SICKLE-CELL DISEASE

*Preparation*

The foundation of our modern knowledge of sickle-cell anemia is based on the research of Dr. Angela Ferguson and Dr. Roland Scott, professors at Howard University in Washington, D.C. The doctors became interested in sickle-cell anemia because many of their friends and some family members were affected by the disease. They published the first research paper on sickle-cell disease in the 1940s—**25 years ahead** of other researchers.

Imagine that you are a physician studying blood genetics in the laboratory of Drs. Ferguson and Scott. The Minister of Health from Zaire has requested their help to investigate two health problems within the villages of his country that seem to be related: sickle-cell disease and malaria. Drs. Ferguson and Scott send you to investigate.

After spending a year looking into the problem, you have collected data about **Family A** and constructed a **pedigree and a health profile** for this family. **Each individual on the pedigree is identified by a number.**

**Exercise 17** ■ *Useful Applications of Genetics* 327

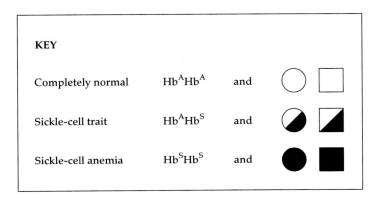

**FIGURE 17-3.** Pedigree of Family A

| TABLE 17-2 |   |
|---|---|
| **HEALTH PROFILE OF FAMILY A** | |
| HEALTH STATUS | AFFECTED FAMILY MEMBERS |
| Affected with malaria | 13, 32, 33 |
| Dead from malaria | 7, 9, 10, 16, 29, 30, 31, 34 |
| Dead from sickle-cell anemia | 8, 15, 25, 26, 35 |
| Dead from causes unrelated to malaria or sickle-cell anemia | 3, 19 |
| In good health | 1, 2, 4, 5, 6, 11, 12, 14, 17, 18, 20, 21, 22, 23, 24, 27, 28, 36, 37, 38 |

**Malaria** is one of the world's most serious diseases. It is transmitted by the bite of an infected mosquito. The bite allows a parasitic protozoan to be passed from the mosquito to the bloodstream of humans. Once in the body, the parasites live and multiply inside red blood cells.

Every two to three days, infected red blood cells burst, releasing thousands of new parasites that will infect even more red blood cells. The infected person experiences bouts of chills and fever each time such a release occurs.

1. Refer to the Health Profile of Family A in Table 17-2. On the **pedigree chart in Figure 17-3, place an asterisk** (*) next to the numbers of all persons who are affected with malaria or who have died from malaria. **List** the pedigree number for **each** individual in Family A who has **died from malaria or is currently affected** with this disease in **Table 17-3**.

| TABLE 17-3 | |
|---|---|
| **GENOTYPES OF INDIVIDUALS AFFECTED BY MALARIA** | |
| PEDIGREE NUMBER FOR EACH AFFECTED INDIVIDUAL (SEPARATED BY GENOTYPE) | GENOTYPE |
| | $Hb^A Hb^A$ |
| | $Hb^A Hb^S$ |
| | $Hb^S Hb^S$ |

**Exercise 17** ■ *Useful Applications of Genetics*

2. **Of the total number of deaths in this family**, what **percentage** died from each cause? Write your answers in Table 17-4.

| TABLE 17-4 CAUSES OF DEATH IN FAMILY A | | |
|---|---|---|
| CAUSE OF DEATH | NUMBER DEAD | PERCENTAGE OF TOTAL DEATHS |
| Malaria | | |
| Sickle-cell anemia | | |
| Other | | |
| TOTAL | | |

3. What is the **leading cause of death** among the members of Family A?

4. **Go back to the pedigree** and examine the genotypes of the individuals who are alive and in good health.

   Among these healthy individuals, what is the most frequent **genotype** for sickle-cell disease? _____

5. **Write a hypothesis** about the relationship between malaria and **one of the three genotypes** related to sickle-cell disease.

6. As you know, a hypothesis is an **educated** guess about a relationship. **Explain** how you **developed** your hypothesis by **listing facts** from the pedigree of Family A.

7. Is the death pattern shown by Family A typical for the population as a whole? To answer this question, **use Figure 17-4 to make a graph** comparing:

a. the percentage of deaths from each cause in Family A (as shown in Table 17-4)

b. the percentage of deaths from each cause in several neighboring communities (as shown in Table 17-5)

### TABLE 17-5
### CAUSES OF DEATH IN NEIGHBORING COMMUNITIES

| CAUSE OF DEATH | NUMBER DEAD | PERCENTAGE OF TOTAL DEATHS |
|---|---|---|
| Malaria | 1080 | |
| Sickle-cell anemia | 600 | |
| Other | 320 | |
| TOTAL | 2000 | |

**FIGURE 17-4.** _____

**Check your graph with your instructor before you continue.**

# Exercise 17 ■ *Useful Applications of Genetics*

## ✓ Comprehension Check

1. In Family A, **family members 21 and 22** are expecting their ninth child. What is the probability that this child will have sickle-cell anemia? **Show your work!**

   |   |   |
   |---|---|
   |   |   |
   |   |   |

2. Closer to home, Michelle had a brother, Charles, who died of sickle-cell anemia. She is concerned about the chance of the condition appearing in her children. When blood samples were taken and placed under low-oxygen conditions, some of her red blood cells sickled. Those of her husband James, however, remained normal when tested. **Show your work and list the genotypes of all those mentioned in the problem.**

   Michelle _____ Charles _____ James _____

   What is the probability that Michelle's and James's children will have sickle-cell anemia? _____

   Sickle-cell trait? _____

**Check your answers with your instructor before you continue.**

Self Test

1. The ability to taste PTC is due to a **dominant allele (T)**. A woman nontaster married a man who was a taster. They had three children. Their two sons were tasters, but their daughter was a nontaster. All four grandparents were tasters.

   **What are the genotypes** of all the individuals mentioned?

   The woman _____        Her husband _____

   The two sons _____     The daughter _____

   Grandparents _____  _____

2. Your sister died from Tay-Sachs disease, inherited as a **recessive** allele **(d)**. You are married and planning to start your family. You are worried about the disease and decide to have genetic testing to see if you or your spouse is a carrier of the Tay-Sachs allele. The test results show that you are a carrier of the allele, but your spouse is not.

   What is the probability that you and your spouse will have a child with Tay-Sachs disease? **Show your work!**

**Exercise 17** ■ *Useful Applications of Genetics*

3. Red-green color blindness is inherited through an **X-linked, recessive** allele (b). Two parents, Fred and Ginger, have normal vision. They have two daughters, Takiyah and Kelly, who also have normal vision, and a color-blind son, David.

   Daughter Kelly has a color-blind son, Kevin. Daughter Takiyah has five sons, all with normal vision. What are the genotypes of all the individuals? **Show all your work!**

   Fred _____ Ginger _____ David _____

   Takiyah _____ Kelly _____ Kevin _____

   Takiyah's five sons _____

   If Kelly marries a man with normal vision, what is the probability that she will have a color-blind son? _____ a color-blind daughter? _____

4. Ralph has normal blood clotting, but he has two brothers and a sister who have hemophilia (an **X-linked, recessive** disorder). What are the most probable genotypes of Ralph's parents? **Explain your answer.**

5. Albinism in humans is expressed as the **absence** of pigment from the skin, hair, and eyes. Using the information in **Figure 17-5**, determine whether albinism is inherited as a **dominant or as a recessive** trait. **Affected** individuals are represented by **shaded** squares and circles.

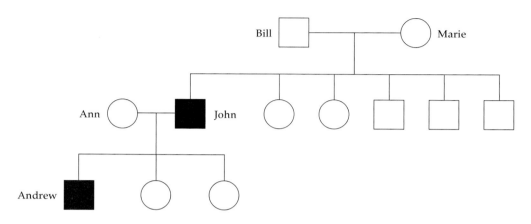

**FIGURE 17-5.** Pedigree Showing Albino Individuals

a. Underneath **each circle and square** in the pedigree, **enter the genotype** of that individual.

b. **(Circle one answer.)** Albinism is a trait that is probably inherited through a **dominant/recessive allele**.

Support your answer with information from the pedigree.

EXERCISE **18**

# Introduction to Molecular Genetics

## Objectives

After completing this exercise, you should be able to:
- demonstrate an understanding of DNA structure and the base-pairing rule
- explain the roles of all of the following in protein synthesis: DNA, mRNA, tRNA, and ribosomes
- complete the processes of transcription and translation from a strand of DNA and determine the sequence of amino acids in the resulting polypeptide chain
- explain how changes in the DNA code may affect protein synthesis and cause health problems.

## CONTENT FOCUS

You have just spent several weeks thinking about genetics and the different ways in which traits can be transmitted from parents to children. Each chromosome is divided into sections called **genes** that are the basis of inheritance. The traits you inherited through your genes (such as hair color, blood type, presence of sickle-cell anemia, or ability to produce the hormone insulin) are all controlled by the production of proteins. **Genes contain the coded instructions your body uses to assemble the hundreds of different types of proteins that make you a unique individual.**

Genes are composed of a molecule known as **deoxyribonucleic acid (DNA)**. In this exercise, you will take a closer look at the **DNA molecule** and the role it plays in all the cells of your body.

## ACTIVITY 1 — REMOVING DNA FROM CELLS

> **Note:**
> Read these directions COMPLETELY before proceeding.

1. **Work in groups.** Get the following supplies: one piece of **calf thymus** (5 grams) in **a dish**, one **graduated 10-ml pipette with manual dispenser**, one **50-ml test tube**, a **test tube rack**, and a pair of **scissors**.

2. On your laboratory table, you will find a **blender, a piece of cheesecloth, a 500-ml graduated cylinder, a container of 6% saline solution, rubber bands, and a 600-ml beaker**.

   Using **scissors, mince** the piece of thymus **as much as possible** and place the pieces in the dish.

   Add 100 ml of **6% saline solution** to the bowl.

> **Note:**
> All groups at your table will be using the blender at the same time, so you MUST coordinate your activities!

3. Coordinating with your partner groups, pour the saline containing the **minced thymus** from **each of the groups** at your laboratory table into the blender.

4. **Put the lid on the blender.** Blend the thymus mixture **on a low setting** for several minutes until no large pieces remain.

5. **Remove the lid** from the blender.

   Assemble **eight thicknesses of cheesecloth** (four pieces from the package, which is manufactured in double thickness). Place the pieces of cheesecloth over the top of the blender and **secure** the cloth **tightly** with the rubber band.

6. Making sure the cheesecloth is **tightly** fastened on the top of the blender, **pour the contents (filtered extract)** into the **600-ml beaker**.

> **Note:**
> At this point, you will go back into your ORIGINAL groups and complete the experiment.

**Exercise 18** ■ *Introduction to Molecular Genetics*

7. Using a **graduated 10-ml pipette with manual dispenser**, remove 10 ml of **filtered extract** from the beaker and place the extract into a **50-ml test tube**. Place the test tube in the rack.

   Place your **used pipettes** in the waste container.

8. Get **one graduated 2-ml pipette** with **manual dispenser, one graduated 10-ml pipette**, and a **thin glass rod**.

   In your supply area, you will find a container of **10% SDS solution** (sodium dodecyl sulfate) and **an ice bucket containing 15-ml screw-cap tubes of 95% ethanol**.

9. To the **test tube containing the filtered thymus** extract, add **1 ml of SDS solution**. Tap the test tube several times to **gently** mix the contents.

> **Note:**
> Read the following directions COMPLETELY before proceeding.

10. Select a **tube of ethanol** from the ice bucket. Using a **clean**, graduated 10-ml pipette, add **10 ml of ice-cold 95% ethanol** to the test tube.

    **With the test tube in the rack**, place the filled pipette with its tip against the **inside wall** of the test tube. **SLOWLY** allow the ethanol to dribble down the inside of the tube, as **demonstrated in Figure 18-1. Don't shake the test tube during this procedure.**

**FIGURE 18-1.** Method for Adding Ethanol to Filtered Thymus Extract

11. The ethanol is lighter than the contents of the tube. When added according to directions, the ethanol will form a **clear layer ABOVE** the filtered thymus extract.

**FIGURE 18-2.** Ethanol Layered above Filtered Thymus Extract

12. Observe the test tube for **five minutes**. The DNA will gradually separate from the thymus mixture and rise into the ethanol layer.

    **Describe** the appearance of the DNA.

    Strand of Bubbles; doesn't really move

13. Let the DNA sample remain undisturbed while you complete the following **Comprehension Check** and **Activity 2**. When you have completed Activity 2, **observe** your DNA sample and see if any changes have occurred. **Record your observations** below.

**Exercise 18** ■ *Introduction to Molecular Genetics*

14. To remove the accumulated DNA from the test tube, follow the directions for **DNA spooling** below:

    a. **Gently insert** the glass rod through **the ethanol layer** into the **accumulated DNA**.

    b. **Carefully twirl** the rod between your fingers, winding the DNA onto the rod, imitating thread on a spool.

    c. **Slowly remove the rod** from the test tube.

## Comprehension Check

1. The thymus is an organ made up of many different types of cells. What percentage of thymus cells contain DNA?

   a. 0%
   b. 25%
   c. 50%
   d. 75%
   e. 100%  *(circled)*

   **Explain your answer.**
   All cells contain DNA.

2. If you removed DNA from a sample of human thymus, how many separate DNA molecules would be removed from **each cell**? 46

3. **(Circle one answer.)** If you repeated the same experiment with an equal number of kidney cells, the amount of DNA collected would **increase/decrease/stay the same**.

   **Explain your answer.**

4. **(Circle one answer.)** If you repeated the same experiment with an equal number of sperm cells, the amount of DNA collected would **increase/decrease/stay the same**.

   **Explain your answer.**

5. *Challenge Question!* **(Circle one answer.)** If you repeated the same experiment with an equal number of red blood cells, the amount of DNA collected would **increase/decrease/stay the same**.

   **Explain your answer.**

   B/c they do not have a nucleus

**Check your answers with your instructor before you continue.**

**Exercise 18** ■ *Introduction to Molecular Genetics*

# ACTIVITY 2                      THE BASICS OF DNA STRUCTURE

Preparation

The extraction of DNA that you just performed showed that DNA is present in cells. It does not, however, give you much information about its actual structure.

DNA molecules are composed of small building blocks called **nucleotides**.

1. Each DNA nucleotide is composed of three smaller molecules hooked together:

   one five-carbon sugar (deoxyribose)
   one phosphate
   one nitrogen base

2. Four different types of nucleotides are needed to build a DNA molecule.

   Each of these four nucleotides has a **different nitrogen base: adenine, guanine, cytosine, or thymine.**

3. DNA has a structure similar to a ladder. The two sides of the ladder are composed of **alternating sugar and phosphate molecules**.

4. Each sugar molecule is attached to **one nitrogen base**. The two strands of DNA are attached by bonds between the nitrogen bases on each side of the ladder.

   Each nucleotide base only bonds with **one specific partner**. The combination of two bases is called a **base pair**.

   **Adenine always bonds with thymine. A-T**

   **Guanine always bonds with cytosine. G-C**

5. Fill in the blanks on the incomplete DNA molecule in **Figure 18-3**.

   Use the symbol **"P"** for **phosphate**, **"S"** for **sugar**, and **"A," "T," "C," or "G"** for the **appropriate nitrogen bases**.

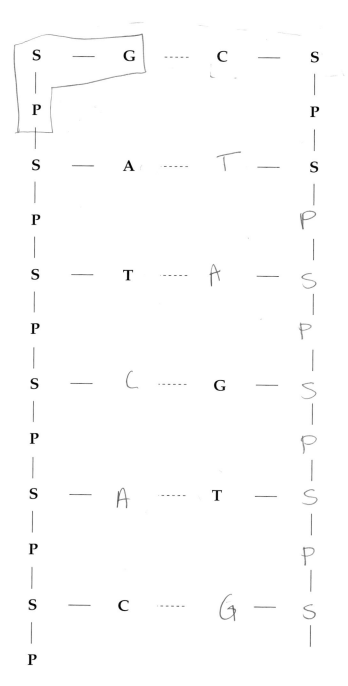

**FIGURE 18-3.** Incomplete DNA Molecule

# Exercise 18 ■ *Introduction to Molecular Genetics*

## ✓ Comprehension Check

1. On **Figure 18-3**, draw a box around one complete nucleotide.

2. **How many nucleotides** are shown in the DNA molecule in **Figure 18-3?** 12

3. How many **different** types of nucleotides were used to construct the DNA molecule in **Figure 18-3?** 4

**Check your answers with your instructor before you continue.**

## ACTIVITY 3 — BUILDING A MODEL OF DNA

1. Work in groups. Get a **DNA-model kit**.

2. Construct the **support stand for the model** by assembling:

   **one gray tube (eight inches long)**

   **three green tubes (two inches long)**

   **one black connector with four prongs**

   Set the stand aside for later use.

3. Separate the parts of the DNA model according to the descriptions in **Table 18-1**.

### TABLE 18-1
### PARTS OF THE DNA MODEL

| Parts of the DNA Molecule | Description of Model Part |
|---|---|
| Deoxyribose sugar | Black, three prongs |
| Phosphate | Red, two prongs |
| Sugar-to-phosphate connectors | Yellow tube |
| Adenine (A) base | Blue tube |
| Thymine (T) base | Red tube |
| Guanine (G) base | Green tube |
| Cytosine (C) base | Gray tube |
| Base-to-base connectors | White, two prongs |

4. Using your knowledge of DNA structure, **assemble the DNA model**. Use the same **base-pairing sequence** as that shown in **Figure 18-3**.

5. Place the model onto the stand by inserting the vertical tube on the stand through the holes in the **white base-to-base** connectors.

   Twist the model as you slide it onto the stand. Stop twisting when the entire molecule fits onto the stand.

   The twisted shape of the DNA molecule is known as a **double helix**.

## Comprehension Check

1. How many nucleotides are present in your DNA model? _____

2. What is the base-pairing rule?

**Exercise 18** ■ *Introduction to Molecular Genetics*

3. Why would it be appropriate to call a DNA molecule a polynucleotide?

4. Assume that the model you just built is an exact representation of **your** DNA code.

    a. Would you use the same bases to construct your lab partner's DNA? _____

    b. Would you assemble the bases in the same order to make a model of your lab partner's DNA? _____

    **Explain your answer.**

**Check your answers with your instructor before you continue.**

Each gene has the instructions to make a single polypeptide chain. These instructions are part of the genetic code. Polypeptide chains are the structural units of proteins. A polypeptide chain is made of many amino acids hooked together.

The key to the genetic code is the sequence of nitrogen bases along **one side** of the DNA molecule. To construct a protein, you must know the **order of the bases**. The code is written in **three-letter "words."** Each of these words (called **triplets**) tells the cell which amino acid should come next when building a protein.

**For proteins to function properly, the amino acids must be assembled in the correct order.**

5. How many **triplets** are present along one side of your DNA model? _____

6. How many **amino acids** will be present in the protein made from your model? _____

> **Note:**
> DON'T take your DNA model apart yet.

# ACTIVITY 4  STEPS OF PROTEIN SYNTHESIS

Preparation

When a particular protein is needed by the body, regions of the double helix unwind so that a cell gains access to the genes that contain the coded information to make that protein. Protein synthesis has two steps: **transcription** takes place in the nucleus and **translation** occurs in the cytoplasm. Both steps require molecules of **RNA (ribonucleic acid)**.

Although the **nucleus** contains the **instructions** for protein synthesis, the **machinery to make proteins** is located in the **cytoplasm**. The coded information is **transferred** from the nucleus to the cytoplasm during **transcription**.

Transcription

1. During transcription, DNA bases are **copied** to form a single strand of RNA, called **messenger RNA (mRNA)**. As with the DNA, mRNA is divided into coded three-letter words. In mRNA, these words are called **codons**.

2. The **base-pairing rule** is used to form messenger RNA, **with one exception**. RNA molecules do not have the nitrogen base thymine. They have **uracil** instead.

Base-pairing to form mRNA:

| DNA BASE | MRNA BASES |
|---|---|
| C | G |
| G | C |
| T | A |
| A | U |

3. The coded information to make a protein appears along **one side** of the double helix. Practice transcription by filling in the correct messenger RNA codons in **Table 18-2**.

### TABLE 18-2
### TRANSCRIPTION OF MRNA

| DNA triplets | CGC | ATA | GAC | TTT | CTT | ACT | TAG | CAT | AAA |
|---|---|---|---|---|---|---|---|---|---|
| mRNA codons | GCG | UAU | CUG | AAA | GAA | UGA | AUC | GUA | UUU |

**Exercise 18** ■ *Introduction to Molecular Genetics*   **347**

4. How many **amino acids** would be in this protein? __9__

5. Which of the five types of nitrogen bases is **not** found in mRNA? _____

Translation

1. A cell needs **amino acids** to construct proteins. The amino acids are carried to the ribosomes by another type of RNA molecule, called **transfer RNA (tRNA)**.

   A tRNA molecule has two functional ends.

   One end picks up amino acids in the cytoplasm (see Figure 18-4).

2. The other end is called the **anticodon**. It contains **three nitrogen bases** that can form a base pair with a **matching codon** in the messenger RNA.

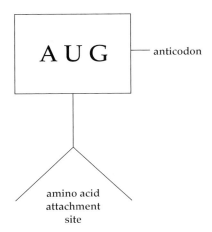

**FIGURE 18-4.** Structure of a Transfer RNA Molecule

3. Each type of tRNA can carry **only one type of amino acid**.

   There are enough different types of transfer RNA molecules to carry all the different types of amino acids needed to make your body's proteins.

4. **Where do the transfer RNA molecules take the amino acids?**

   They take them to **ribosomes**, organelles in the cytoplasm where proteins are manufactured.

Ribosomes are made of proteins and a third type of RNA called **ribosomal RNA**.

5. **Ribosomes read messenger RNA codons and accept amino acids brought by transfer RNA molecules**.

   Ribosomes hook amino acids together in the **order specified by the messenger RNA** codons to construct the polypeptide chain.

Summary of Protein Synthesis

- DNA contains the instructions to make polypeptide chains. A region of the DNA double helix unwinds. Coded instructions to make a protein are exposed.

- This information is carried to the cytoplasm by messenger RNA molecules (transcription).

- Amino acids in the cytoplasm are used to build polypeptides.

- Transfer RNA molecules pick up the amino acids and transport them to ribosomes, the locations where proteins are made.

- Ribosomes bond amino acids together according to the instructions in the genetic code.

**Exercise 18** ■ *Introduction to Molecular Genetics*   349

# ACTIVITY 5                                        BUILDING A REAL PROTEIN

Preparation

Imagine the following situation: you are about to give birth to a baby. The brain produces the hormone **oxytocin** (a small protein), which causes uterine muscles to contract for childbirth. After birth, this same hormone causes muscles in the mammary glands to contract, releasing milk for nursing the baby.

Suppose this is your first baby. How does the brain know how to manufacture oxytocin if it has never been needed before? The information is stored in your DNA *"reference"* library.

1. Using the steps outlined in **Activity 4** as guidelines, build the protein **oxytocin**.

   Oxytocin is one of the smallest proteins, with only nine amino acids connected into a single polypeptide chain.

2. **Transcribe** the DNA triplets that code for oxytocin into messenger RNA codons and add them to the appropriate spaces below.

3. The messenger RNA must detach from the DNA and leave the nucleus for **translation** to take place.

   Cut out the strip of mRNA and move it to the ribosome in **Figure 18-5**. Notice that the ribosome has two parts, separated by a groove. The mRNA should be placed along the groove between the upper and the lower sections.

4. Each transfer RNA molecule in **Figure 18-6** is ready to deliver an amino acid to the ribosome.

   Where should each transfer RNA molecule deliver its amino acid?

   Cut out the **transfer RNA** molecules.

   Place each tRNA molecule under the correct mRNA codon.

> **Hint:**
> You will be able to match each tRNA with its codon by following the base-pairing rule.

| DNA TRIPLETS | ACG | ATG | TAT | GTT | TTG | ACG | GGA | GAC | CCC |
|---|---|---|---|---|---|---|---|---|---|

| MRNA CODONS | UGC | UAC | AUA | CAA | AAC | UGC | CCU | CUG | GGG |
|---|---|---|---|---|---|---|---|---|---|

5. Refer to **Table 18-3** to determine the name of each amino acid in your oxytocin polypeptide chain.

   Each box at the bottom of **Figure 18-5** represents **one amino acid** in the chain.

   **Write the names** of the amino acids in order in each box.

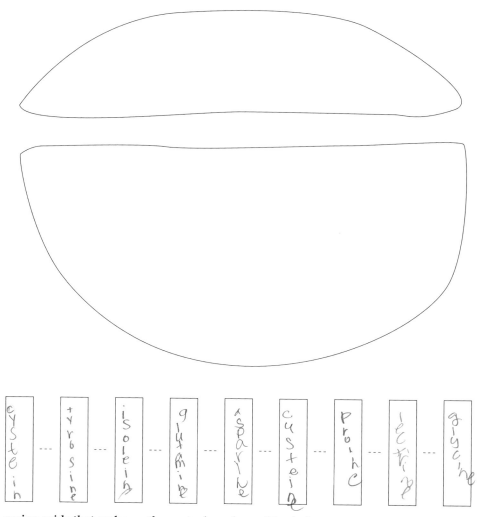

amino acids that make up the oxytocin polypeptide chain

**FIGURE 18-5.** Translation

 **Check your oxytocin molecule with your instructor before you continue.**

### Exercise 18 ■ Introduction to Molecular Genetics

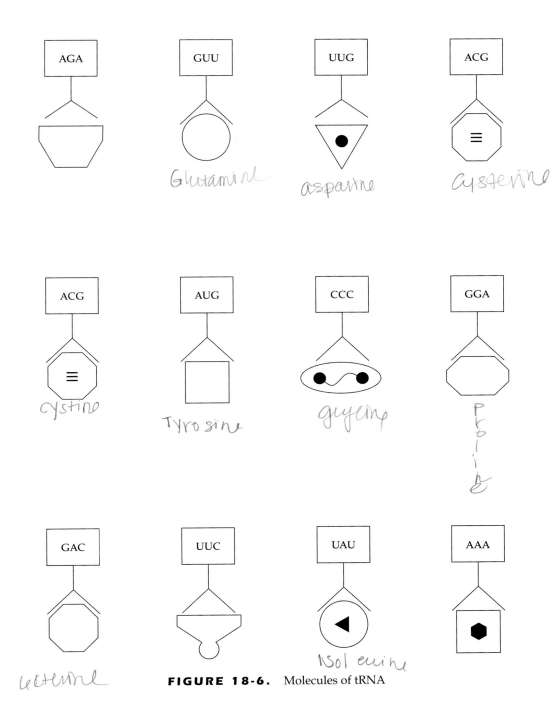

**FIGURE 18-6.** Molecules of tRNA

**Exercise 18** ■ *Introduction to Molecular Genetics*

### TABLE 18-3
### KEY TO IDENTIFYING AMINO ACIDS

| Shape | Name of Amino Acid |
|---|---|
| (pentagon/home-plate) | Histidine |
| (circle) | Glutamine |
| (triangle with dot) | Asparagine |
| (octagon with ≡) | Cysteine |
| (square) | Tyrosine |
| (oval with two dots connected) | Glycine |
| (elongated octagon) | Proline |
| (hexagon) | Leucine |
| (flask shape) | Alanine |
| (circle with left triangle) | Isoleucine |
| (square with small hexagon) | Serine |

## Comprehension Check

1. If the second amino acid in your polypeptide chain were **valine**, would the protein still be oxytocin? _____ **Explain your answer**.

2. **(Circle one answer.)** The DNA triplet A A A would be transcribed into the mRNA codon **T T T/U U U**.

3. Put the following steps of protein synthesis in the correct order.

   _____ tRNA molecules pick up amino acids

   _____ mRNA transcribed

   _____ DNA double helix unwinds

   _____ mRNA binds to ribosome

   _____ Ribosome bonds amino acids together

   _____ tRNA anticodon links with mRNA codon

   _____ mRNA leaves nucleus

   _____ Polypeptide chain completed

**Check your answers with your instructor before you continue.**

EXERCISE 19

# Biotechnology: DNA Analysis

## Objectives
After completing this exercise, you should be able to:
- discuss the differences between genes and non-coding regions of DNA
- summarize how the process of gel electrophoresis separates DNA molecules
- explain the role of non-coding DNA regions in producing a DNA profile
- draw conclusions that are supported by analysis of RFLP and STR profiles
- interpret DNA data presented in the form of tables, charts, and/or graphs
- explain how STR analysis can be used to determine paternity
- apply your knowledge of DNA fingerprinting to real-life situations.

## CONTENT FOCUS

Half of your 46 chromosomes were inherited from your mother and the other half from your father. Consequently, **everyone has a unique DNA pattern** (except for identical twins). Tissue samples containing DNA can be used for identification in criminal cases, in paternity suits, and in cases where visual identification is not possible. The technique used to make these genetic comparisons produces a **DNA profile (also called a DNA fingerprint).**

DNA for examination is isolated with a process similar to the method you used in the DNA spooling laboratory. As you know, DNA is located in the nucleus of all body cells. Therefore, DNA can be extracted from any tissue. Common sources include **white blood cells, hair roots, semen, saliva** (which contains **epithelial cells**), and other body tissues.

So far, you have studied basic genetics and the structure of the DNA molecule. You have seen that various traits are produced on the basis of your genetic code (DNA). Not all of a person's DNA codes for the synthesis of proteins, however. **DNA molecules also have areas that are not genes.**

Originally, this non-coding DNA was called "**junk DNA**" because scientists did not understand or recognize the function for these sections of the chromosomes (see **Figure 19-1**). They originally believed that these sequences were present to simply fill the gaps between the genes. Although the functions for most non-coding regions have not yet been discovered, scientists are beginning to find evidence that some of these repetitive DNA sequences play important roles in cellular metabolism and inherited diseases.

Gene 1                      Junk DNA                      Gene 2

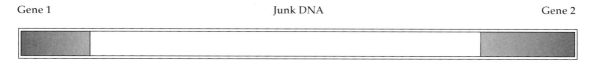

**FIGURE 19-1.** Position of Genes and Non-coding Regions of DNA

Regions of non-coding DNA vary a great deal among individuals, and when specific DNA regions are studied, scientists can use this information to establish human identity, analyze evolutionary trends, and determine predispositions to certain diseases.

The pattern of non-coding regions between any given two genes is unique and always has the same repeating pattern of nucleotides (for example **C A T**). The number of times these nucleotides are repeated, however, is highly variable among people.

As you can see in **Figure 19-2**, Person A has nine repeating **C A T** segments, while Person B has only five. Some people have dozens of repeats of the same pattern.

Each person has a distinct **DNA profile** (or fingerprint).

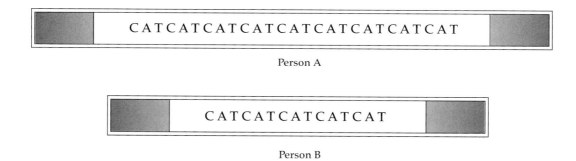

**FIGURE 19-2.** Repeating Non-coding DNA Segments

One of the original processes used to create a DNA profile is called **RFLP Analysis**. RFLP stands for **restriction fragment length polymorphism**. DNA extracted from a blood or a tissue sample is cut into small pieces at specific locations using special chemicals called **restriction enzymes**.

**Fragment length polymorphism** refers to the fact that the **pieces (fragments) of non-coding regions are different lengths for different people.**

# Exercise 19 ■ Biotechnology: DNA Analysis

To make a reliable "RFLP match," molecular geneticists use several different non-coding DNA regions. When several different regions are compared, the odds against a mistaken match can be **more than a billion to one**.

## ACTIVITY 1 — RFLP FINGERPRINTS

Preparation

An early form of forensic DNA analysis used **restriction enzymes (enzymes that cut the DNA molecule at specific sites)** to produce fragments of DNA. These fragments vary in size from individual to individual based on the number of times that a core sequence is repeated.

Fragments of DNA are separated using a process called **gel electrophoresis**. The **gel** is a thin sheet of gelatin supported by a glass plate with **electrodes** attached at both ends. One end of the gel has a **positive** charge and the other end has a **negative** charge.

The DNA fragments are attracted to the **positive end** of the gel plate. **Smaller, lighter** fragments migrate more easily through the gel and therefore travel farther in a given time period than larger, **heavier** fragments. The light and heavy fragments form bands across the gel layer (see **Figure 19-3**).

The DNA bands of interest are **marked** with special **molecular probes**. Probes are small pieces of DNA that use the **base-pairing rule** to locate and bind **only** to the fragments that will be used to form the DNA profile.

**FIGURE 19-3.** Gel Electrophoresis of DNA

The sheet of DNA bands is "photographed" with **X-ray film**. The **only** bands visible on the developed film are those that have been **labeled with molecular probes**. The X-ray **pattern of bands** for several DNA locations is analyzed to create a **DNA profile** for an individual (see **Figure 19-4**, which contains samples of several DNA profiles). Each profile is represented by a **vertical column of DNA bands**.

For RFLP testing, **several different regions (called loci) of junk DNA are analyzed**. If **ANY** of the loci **do not match** between an evidence sample and a known individual, **the person is eliminated** as a possible source of the evidence DNA.

If, however, **all of the loci in the evidence sample** exhibit the **same pattern as the known** sample, calculations are conducted to determine how rare or common the profile is.

**FIGURE 19-4.** Sample DNA Fingerprints

## Comprehension Check

**Figure 19-5** shows some samples of DNA profiles produced by multiple tests of RFLP loci. Each profile is represented by a **vertical column of DNA bands**.

Two profiles match and two do **not** match. Matching profiles must have **exactly** the same banding patterns. The bands have been numbered for reference.

# Exercise 19 ■ Biotechnology: DNA Analysis

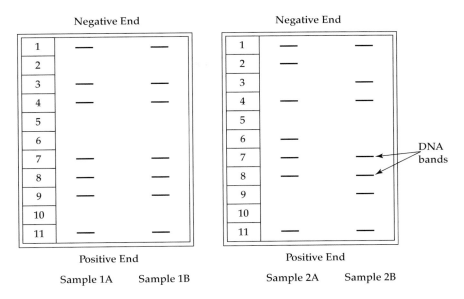

**FIGURE 19-5.** Four DNA Profiles

1. (**Circle one answer.**) The DNA sample was placed in the gel at the **negative/positive** end of the tray.

2. List the bands that **don't match** in **samples 2A and 2B**. _____

3. Which band in each sample contains the **heaviest** DNA fragments? _____

4. Which band contains the **lightest** fragments? _____

5. How many DNA bands are present in **sample 2B**? _____

6. (**Circle one answer.**) If the DNA fragments are migrating toward the positive end of the gel, DNA molecules must have a **negative/positive** charge.

7. *Challenge Question!* What could you conclude from the RFLP analysis if you knew that the DNA in **sample 1A** was from blood and the DNA in **sample 1B** was from skin?

**Check your answers with your instructor before you continue.**

# ACTIVITY 2   SOLVING A CRIME USING RFLP FINGERPRINTS

*Preparation*

On January 1, at about 2:00 a.m., the police responded to a report of gunshots at the 600 block of Knorr Street. Arriving at the scene, the officers found a parked Dodge pickup. In the front seat was a deceased male with several gunshot wounds.

Shoe marks were visible in a pool of blood outside the truck, but they were not clear enough to identify the type of footwear that made them. Homicide detectives have narrowed the field to three suspects.

**Suspect A** was arrested near the scene. He was wearing black work boots that appeared to have dried blood on the soles.

At the home of **Suspect B**, investigators recovered a pair of tennis shoes that also appear to have dried blood on the soles. Suspect B says she had a severe nosebleed, which might have resulted in blood on her shoes.

**Suspect C** was the roommate of the deceased. When officers went to the apartment to notify him of the death, Suspect C was packing a suitcase. He was wearing high-top sneakers that appeared to have dried blood on the soles and sides. When questioned about the blood and the murder, Suspect C had no comment.

All the samples were tested and identified as human blood. The homicide detectives have asked the Central Police DNA Laboratory to analyze the blood evidence using RFLP analysis. You are the forensic scientist assigned to the case. **Figure 19-6** shows the DNA profiles from all the blood samples.

1. Note that the box under the **first column** in **Figure 19-6** (blood of victim) contains the number **1**. This is the first unique DNA profile found.

2. Moving from left to right, examine each DNA profile.

   If the new profile **does not match any previously examined profile**, enter a **new number** in the box under the column (**2, 3, 4,** etc.).

   If the new DNA profile **is an exact match** of a previous profile, enter the **number you gave to the profile that matches this new profile**.

   Continue with this process until you have examined all the DNA profiles in **Figure 19-6**.

## Exercise 19 ■ Biotechnology: DNA Analysis

| | Blood of Victim | Blood on Ground at Crime Scene | Blood of Suspect A | Blood of Suspect B | Blood of Suspect C | Blood from Suspect A's Shoes | Blood from Suspect B's Shoes | Blood from Suspect C's Shoes |
|---|---|---|---|---|---|---|---|---|
| 1 | | | — | | — | — | | |
| 2 | — | — | | | | | | — |
| 3 | — | — | | | — | | | — |
| 4 | | | — | | | — | | |
| 5 | | | — | — | | — | — | |
| 6 | | | | — | — | | — | |
| 7 | — | — | | — | | | — | |
| 8 | | | | | — | | | |
| 9 | | | — | | — | — | | |
| 10 | — | — | | | | | | — |
| 11 | | | | — | | | — | |
| 12 | | | — | | — | — | | |
| 13 | | | | — | | | — | |
| 14 | | | | | | | | |
| 15 | — | — | | | | | | — |
| 16 | — | — | | | | | | — |
| 17 | | | | | | | | |
| 18 | | | | — | | | — | |
| 19 | | | | | | | | |
| 20 | — | — | — | | | — | | — |

| | 1 | | | | | | | |

**FIGURE 19-6.** DNA Profiles—Knorr Street Homicide

## ✓ Comprehension Check

1. How many **unique** profiles were found? _____

2. Did any profiles match the blood from **Suspect A's shoes**? _____ If so, list them:

3. Did any profiles match the blood from **Suspect B's shoes**? _____ If so, list them:

4. Did any profiles match the blood from **Suspect C's shoes**? _____ If so, list them:

5. Did any profiles match the **blood from the deceased male**? _____ If so, list them:

6. Did any profiles match the **blood found on the ground** outside the truck? _____ If so, list them:

7. Do these **results** make any individual a more likely suspect than the others? _____

   **Explain your answer**, citing facts from the DNA analysis.

8. What other types of evidence would be helpful to be sure you had found the murderer?

**Check your answers with your instructor before you continue.**

Exercise 19 ■ *Biotechnology: DNA Analysis*

# ACTIVITY 3    SHORT TANDEM REPEAT (STR) ANALYSIS

Frequently, the amount of DNA collected at crime scenes is insufficient for RFLP analysis. For this reason, forensic scientists began to use a procedure called the **polymerase chain reaction (PCR)**. PCR allows the scientist to **target specific chromosome locations and make copies of the DNA**. By using this technique, scientists can copy samples containing a small amount of DNA (for example, residue left on the back of a licked envelope.)

Scientists currently use non-coding **DNA loci** called **short tandem repeats (STRs)** in forensic DNA testing. **An STR is a small sequence of bases that is repeated**. The number of times that the unit is repeated varies from person to person. For example, an STR with the **base sequence** AATC may be repeated between 12 and 24 times.

As with all of your DNA, the alleles for this example STR locus would be inherited from your parents.

Each individual would have **two alleles for this STR locus** (recall that each person inherits one allele from each parent).

So, an individual's **genotype for this locus could be 13, 14 (meaning that the allele this person inherited from one parent repeats the STR base sequence 13 times and the other allele repeats 14 times).**

To make an **STR profile**, the DNA sample is tagged with a **fluorescent label**. The label is detected by a **laser**, which sends a **signal to a computer**. This signal produces a peak on a computer printout.

The pattern of peaks on an STR profile (see **Figure 19-7**) represents the alleles at each locus (recall that there are two alleles at each locus). If you see only **one peak** for a locus, the individual has **two alleles that are alike. They are homozygous** at that locus.

In the sample locus to the right, note that there are **numbers below each peak**.

These two numbers are used to designate the person's **genotype**.

In this example, the **genotype is 13,14**.

The 13 allele has the STR "phrase" AATC repeated 13 times. The 14 allele has the STR phrase AATC repeated 14 times.

The **letter and number above the peaks** are **abbreviations** that represent the different STR loci.

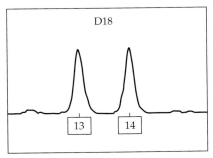

**FIGURE 19-7.** Computer Printout of Sample Alleles at One STR Locus

As with RFLP, an STR profile does not consist of only one locus. **Multiple loci are used to develop a DNA profile.** The forensic community has adopted **13 core STR loci** that are tested by all forensic laboratories. By having all labs perform the same test, results from around the country can be compared. These results are currently being stored and compared in a **computer database system** called **CODIS (Combined DNA Index System).**

DNA samples are frequently analyzed in **two STR sets**. Group 1 contains **nine STR loci** plus one locus, abbreviated **AML**, that is used for **gender identification (XX or XY).**

**Group 2** contains **the remaining four STR loci, plus the gender locus, plus two STR loci from Group 1 that are repeated** as a **control in Group 2**, to ensure that samples originated from the same individual.

A complete STR profile is shown in **Figure 19-8**.

**FIGURE 19-8.** Example of an STR Profile

**Exercise 19** ■ *Biotechnology: DNA Analysis*

## ✓ Comprehension Check

1. **(Circle one answer.)** This DNA profile was taken from a **male/female**. **Explain your answer.**

2. Using the STR abbreviations on the profile, which two loci are the controls (repeated in both Groups 1 and 2)?

3. How many alleles in **Group 1** of the STR profile are homozygous? _____

   How many are heterozygous? _____

   How many alleles in **Group 2** of the STR profile are homozygous? _____

   How many are heterozygous? _____

   **Explain your answers.**

4. What do the numbers 9 and 13 designate in the **D7 locus** of Group 2?

5. *Challenge Question!* From your answer to Question 3, can you tell if any of these alleles are dominant or recessive? **Explain your answer.**

# ACTIVITY 4 — A CRIMINAL CASE USING STR EVIDENCE

The local police department responded to a call regarding a sexual assault. Upon arriving at the scene, they found the victim lying on the floor, partially clothed, and crying. The victim, Jane Doe, was transported to the hospital for treatment and evidence collection. Using a brief description of the suspect, the police apprehended two individuals, but the victim was not able to make a positive identification.

At the hospital, **two evidence samples** were collected from the body of the victim. The evidence samples **tested positive for semen** and were sent to the lab for STR analysis. In addition, **known blood samples from all individuals** were obtained and submitted for STR analysis.

1. Interpret the STR profiles in **Figures 19-9 through 19-13** on the following pages and **write the genotypes in the STR table** below.

| Grp 1 Sample | D3 | vWA | FGA | AML | D8 | D2 | D18 | D5 | D13 | D7 |
|---|---|---|---|---|---|---|---|---|---|---|
| Semen sample one | | | | | | | | | | |
| Semen sample two | | | | | | | | | | |
| Victim | | | | | | | | | | |
| Suspect one | | | | | | | | | | |
| Suspect two | | | | | | | | | | |

| Grp 2 Sample | D3 | D16 | AML | TH01 | TPOX | CSF | D7 |
|---|---|---|---|---|---|---|---|
| Semen sample one | | | | | | | |
| Semen sample two | | | | | | | |
| Victim | | | | | | | |
| Suspect one | | | | | | | |
| Suspect two | | | | | | | |

Exercise 19 ■ *Biotechnology: DNA Analysis* 371

Group 1

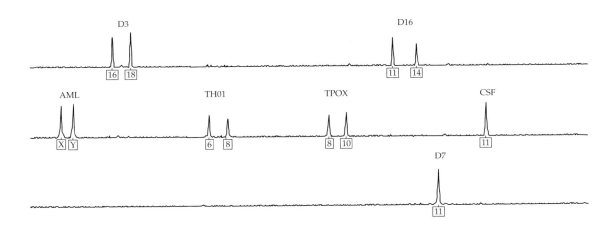

Group 2

**FIGURE 19-9.** STR Profile of Semen Sample One

372     *Lab Manual for Biology of Humans* ■ Exercise 19

Group 1

Group 2

**FIGURE 19-10.** STR Profile of Semen Sample Two

Exercise 19 ■ Biotechnology: DNA Analysis

**FIGURE 19-11.** STR Profile of Known Blood Sample from the Victim

Group 1

Group 2

**FIGURE 19-12.** STR Profile of Known Blood Sample from Suspect One

# Exercise 19 ■ Biotechnology: DNA Analysis

**Group 1**

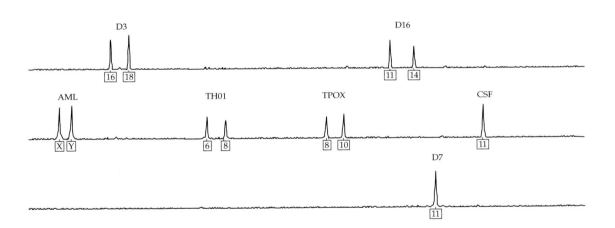

**Group 2**

**FIGURE 19-13.** STR Profile of Known Blood Sample from Suspect Two

## Comprehension Check

1. On the basis of the results of your analysis, what conclusions can you draw about this case? **Be specific. Support your answer with facts** from the STR profiles.

2. The detective investigating this case asks you the following question. Please respond.

   "If the DNA profile from the evidence matches the suspect, he must be guilty of sexual assault, right?"

3. **(Choose one answer.)** When a blood sample is submitted for STR analysis, the DNA tested is taken from **red blood cells/white blood cells/platelets. Explain your answer.**

Exercise 19 ■ *Biotechnology: DNA Analysis* 377

4. What results would you expect from a comparison of STR profiles taken from the following? **Explain your answers**.

   a. Identical twins:

   b. Fraternal twins:

 Check your answers with your instructor before you continue.

## ACTIVITY 5    A PATERNITY SUIT INVOLVING STR ANALYSIS

Preparation

STR profiles can also be used to help identify a child's biological parents. As you recall from studying meiosis, a child inherits half of its genetic information from each parent. This makes matching of DNA profiles more challenging.

**If a child has alleles in his or her STR profile that don't match those of the mother, these alleles must have been inherited from the father.**

A young woman claims that her two-year-old son is the child of one of the members of the hit rock group "First Impressions." During the time of conception, she was dating all four of the band members.

In her sworn testimony, the mother states, "The father is one of the band members. They were the people I spent time with during that whole year. The problem is, I don't know which musician is the biological father of my baby."

None of the members of the band have admitted to fathering the child. During the investigation, all members of the rock group agreed to donate samples of their blood for DNA analysis. You are the forensic scientist appointed by the court to analyze the DNA evidence.

The profiles for all parties involved are summarized in the following STR table.

1. Compare the child's STR profile with the mother's. In the child's profile, **cross out** any **Group 1 and Group 2** alleles that are the same as the mother's (as shown in the example below).

2. Next, compare the alleles of each of the potential fathers with the child's profile. **Circle all matching alleles.**

3. On the basis of your analysis of the STR profiles, which of the men could be the father of the child? **Explain** your answer, using facts from the STR analysis.

 **Check your answers with your instructor before you continue.**

| Grp 1 | D3 | vWA | FGA | AML | D8 | D2 | D18 | D5 | D13 | D7 |
|---|---|---|---|---|---|---|---|---|---|---|
| Child | 13, 18 | 12, 12 | 20, 25 | X, X | 9, 18 | 28, 30 | 10, 12 | 7, 16 | 13, 13 | 6, 9 |
| Jenny Smith | 12, 18 | 12, 15 | 20, 20 | X, X | 9, 18 | 28, 32 | 12, 19 | 12, 16 | 13, 14 | 6, 6 |
| Peter Niles | 13, 13 | 11, 12 | 22, 25 | X, Y | 9, 9 | 31, 32 | 9, 10 | 10, 12 | 8, 9 | 9, 11 |
| Phillip Cruze | 12, 13 | 12, 20 | 25, 29 | X, Y | 10, 18 | 30, 30 | 10, 18 | 7, 9 | 11, 13 | 9, 15 |
| Mike Miller | 13, 19 | 12, 12 | 24, 26 | X, Y | 9, 15 | 31, 31 | 14, 15 | 13, 16 | 15, 15 | 6, 8 |

| Grp 2 | D3 | D16 | AML | TH01 | TPOX | CSF | D7 |
|---|---|---|---|---|---|---|---|
| Child | 13, 18 | 11, 14 | X, X | 9, 11 | 7, 11 | 13, 14 | 6, 9 |
| Jenny Smith | 12, 18 | 8, 11 | X, X | 10, 11 | 7, 13 | 6, 14 | 6, 6 |
| Peter Niles | 13, 13 | 9, 13 | X, Y | 5, 7 | 8, 9 | 12, 14 | 9, 11 |
| Phillip Cruze | 12, 13 | 14, 16 | X, Y | 9, 10 | 11, 13 | 11, 13 | 9, 15 |
| Mike Miller | 13, 19 | 11, 16 | X, Y | 7, 11 | 8, 12 | 9, 10 | 6, 8 |

Exercise 19 ■ *Biotechnology: DNA Analysis*    **379**

Self Test

1. What is meant by "non-coding" DNA regions?

2. How can non-coding DNA regions be used to determine someone's identity? **Be specific**.

3. How is comparing STR profiles to determine paternity different from trying to match a suspect to a blood sample?

4. You are a park ranger at Yellowstone National Park. You have discovered the internal organs of a deer just inside the park boundaries. Because the park is protected from hunting, you notify the local police to be on the alert for a recently killed deer. The police spot a truck carrying the carcass of a large deer a few miles outside the park. They suspect this is the deer that was killed inside the park boundaries, but they need some evidence linking this deer to the remains found in the park.

   What test(s) could determine whether the deer in the truck was killed inside Yellowstone Park? **Explain your answer**.

5. On Saturday, September 28, a young boy was bitten by a large black dog on the corner of Fifth Avenue and Tenth Street. Investigating the situation further, police officers learned that a neighbor has a large black dog that often barks ferociously at people walking down the street. The owner maintained that the dog made a lot of noise, but was basically friendly. The officers, however, discovered a substance, which appeared to be blood, on the dog's collar. The owner stated that he cut his hand on the dog collar and that the blood was his.

The boy's family decides to sue the dog's owner. They wish to be compensated financially and also have the dog destroyed. However, the dog's owner is persistent that his dog never left his fenced yard on the date in question. You are a forensic scientist for a private firm that has been asked to evaluate the case for the civil trial. All samples have been tested and found to be human blood. Interpret the results of the DNA testing as presented in **Figures 19-14 through 19-16**.

Summarize your findings in the dog bite investigation. **Support your statements with evidence** from the DNA fingerprints.

# Exercise 19 ■ Biotechnology: DNA Analysis

**Group 1**

**Group 2**

**FIGURE 19-14.** Dog Owner's Blood

382  Lab Manual for Biology of Humans ■ Exercise 19

Group 1

Group 2

**FIGURE 19-15.** Boy's Blood

## Exercise 19 ■ Biotechnology: DNA Analysis

Group 1

Group 2

**FIGURE 19-16.** Blood from Dog's Collar

EXERCISE 20

# Using Biotechnology to Assess Ecosystem Damage

## Objectives
After completing this exercise, you should be able to:
- explain the effects of environmental factors on bioluminescence and give examples
- correctly prepare and use serial dilutions
- demonstrate awareness of factors that can affect the accuracy of experimental results
- explain the utility of serial dilutions to analyze the effects of chemicals on ecosystems
- use experimental results to rank chemicals in terms of their potential to cause harm to human health or the environment
- draw graphs that present data clearly and accurately
- interpret data in tables, charts, and graphs
- apply your knowledge of chemical toxicity to real-life situations.

## CONTENT FOCUS

Many types of harmful compounds are released into the environment by industry, agriculture, and even from our own households. It is important not only to detect harmful materials but also to predict their effects on the environment. This is difficult because pollutants are often released in **small amounts, which rapidly disperse** into the soil or water. It is also important to identify **which types** of chemicals are present in the environment, and whether these chemicals are **harmful** to living organisms.

To address this problem, Dr. Kenneth Thomulka, a scientist from Philadelphia, has developed an inexpensive, convenient field-testing method to discover whether toxic chemicals are present in the environment.

The method uses harmless marine bacteria as **biological indicators** for the presence of toxic chemicals.

## ACTIVITY 1 — EXPLORING BIOLUMINESCENCE

*Preparation*

The ability of an organism to produce light is called **bioluminescence**. Bioluminescence is exhibited in a variety of organisms, from bacteria to fireflies in your backyard. It is the only light source for marine organisms living deep in the ocean. Like fireflies, some deep-sea fishes use their lights like signals to find mates. Others, like the deep-sea anglerfish, wave glowing lures to attract smaller fish as prey.

Many luminescent organisms, including bacteria, have the enzyme **luciferase**. Luciferase and **oxygen** are needed for the complex chemical reactions used to produce light.

> **Note:**
> Read these instructions COMPLETELY BEFORE the lights go out!

1. Work in groups. Get the following supplies: **a test tube rack and one 15-ml screw-capped test tube containing a sample of the bacterial culture.**

2. Place the test tube in the rack and set it on your laboratory table. **Unscrew and remove the lid.** Leave the test tube **undisturbed** for 15 minutes.

> **Caution:**
> At times during this laboratory period, the room will be completely dark. Make sure the aisles and working areas around your table are clear.

3. To demonstrate light production, the class will observe a flask containing **bioluminescent marine bacteria**. With the **lights out**, observe what happens when the instructor swirls the flask containing the bacteria. **Record** your observations.

4. **Continue to observe** the flask for several minutes as it sits undisturbed on the laboratory counter. **Record** your observations.

**Exercise 20** ■ *Using Biotechnology to Assess Ecosystem Damage*   **387**

5.  **Without moving or disturbing the test tube rack**, check the tube of bacteria on your laboratory table. Observe the tube **very carefully**.

    Where is the light level **most** intense? _____

    Where is the light level **least** intense? _____

    Using the information presented in the **introduction** to this exercise, which mentions **two** substances needed for bioluminescence to occur, **explain** your observations.

6.  **Rank the level of light produced** by the bacteria on a scale of **0 (no light) to 4 (most light)**.

    Top of the tube _____

    Middle of the tube _____

    Bottom of the tube _____

7.  **Shake** the test tube **gently** and observe the results. **Describe** what you see.

8.  After shaking the test tube, **rank the level of light produced** by the bacteria on a scale of **0 (no light) to 4 (most light)**. _____

    **Why** did the change in light level occur?

Bacterial test kits are commercially available for detecting pollutants in aquatic ecosystems, such as lakes, streams, and oceans. Manufacturers of household cleaners, shampoos, and cosmetics have started using bacterial testing to replace the more traditional, but highly controversial, tests done with rabbits, rats, and mice.

In the following experiments, you will use bioluminescent bacteria to evaluate the toxicity of common household products. You will try to determine whether any of these household products can be damaging to the environment if they are **not used and disposed of properly**.

Any environmental condition that is harmful to the bacteria's health will cause a decrease in light production.

The point of this experiment is NOT to see how well you can kill bacteria. These bacteria are not harmful. They are valuable and necessary for the environment. The point is to find out how you can PREVENT household chemicals from harming organisms in the environment, including bacteria.

## ACTIVITY 2 — GETTING STARTED

1. Work in groups. Get the following supplies: **a small flashlight with red filter, one small beaker, one graduated 10-ml pipette with manual dispenser, and one graduated 1-ml pipette with manual dispenser.**

   You will also need: **six large test tubes in a rack, six small test tubes in a rack, scissors, masking tape, a dispenser bottle filled with 3% saline solution, a tray for used glassware, and a container for waste fluids.**

2. Fill the small beaker half full with **tap water**. Attach the manual dispenser to the end of **one graduated 10-ml pipette**.

   **Practice filling and dispensing fluid from the pipette** following the directions given by your instructor.

3. Now that you are an expert pipette user, you are ready to set up your experiment.

   With masking tape, label the **large** test tubes **D1** through **D6**.

   Make sure you place the tape labels as **close as possible to the top of each tube**.

Exercise 20 ■ *Using Biotechnology to Assess Ecosystem Damage*

4. The following household products are available for testing:

   | | |
   |---|---|
   | Drain cleaner | Automobile antifreeze |
   | Mouthwash | Dishwashing liquid |
   | Household cleaner | Herbicide (weed killer) |
   | Toilet-bowl cleaner | Pet shampoo |

   ### Caution:
   **Some of these products may be harmful to your skin or eyes. Be careful not to spill any on your hands as you measure. If a spill occurs, DO NOT touch your face or eyes. Wash your hands thoroughly before proceeding.**

5. Your instructor will assign a household product to your group.

   **Obtain a small container filled with your assigned household product.**

   Enter the name of the chemical your group will be using: _____

6. You will **dilute** the chemical with various amounts of saline to test the reaction of the bacteria to **different strengths** of this product.

   The dilutions will produce a series of proportionally weaker solutions, commonly referred to as **serial dilutions**. In this manner, you can determine what dilution (concentration) of this chemical is harmful to the bacteria.

   **In this experiment, we will compare chemicals to see which are more harmful than others and we will base our assessment of toxicity on the level of dilution necessary to dispose of the chemical safely.**

   Rather than just diluting randomly to make this assessment, it is more useful to dilute the chemical in **measured steps (making a serial dilution)**. By exposing bacteria to different concentrations (dilution levels) of the various household chemicals, you can determine the exact level of dilution that will make the chemical harmless.

   Since these are marine bacteria, you will make your dilutions with **3% saline** (a mild salt solution), which simulates the water in the bacteria's natural environment.

7. The dilution levels you will use are listed in **Table 20-1**.

| TABLE 20-1 DILUTION SEQUENCE | | |
|---|---|---|
| TEST TUBE NUMBER | DILUTION | PERCENT CONCENTRATION OF CHEMICAL |
| D1 | Full strength | 100 |
| D2 | 1:10 | 10 |
| D3 | 1:100 | 1 |
| D4 | 1:1,000 | 0.1 |
| D5 | 1:10,000 | 0.01 |
| D6 | No chemical added | 0 |

# ACTIVITY 3 — MAKING ACCURATE SERIAL DILUTIONS

> **Hint:**
> To obtain valid results, you must be very careful to make accurate measurements.

1. Using the **10-ml graduated pipette**, transfer **9 ml** of your assigned household chemical into the first **large** test tube (marked **D1**). This will be the **full-strength** tube.

> **Note:**
> Place all used pipettes and glassware into the container provided.

2. Using a **1-ml graduated pipette**, transfer **1 ml** of household chemical from the container with the original sample into the next large test tube (marked **D2**).

**Exercise 20** ▪ *Using Biotechnology to Assess Ecosystem Damage*

3. To the **same** tube (**D2**), add **9 ml of saline** solution from the stock bottle. **To dispense the saline**,

   a. hold the test tube under the spout
   b. **slowly** raise the pump handle as far as it will go
   c. **gently** push the handle down to dispense

   Without spilling the contents, **shake the test tube gently** to mix. This will be the **1:10 dilution**.

4. **From test tube D2, remove 1 ml** of solution and transfer it to test tube **D3**.

   Following the instructions above, add **9 ml of saline** to tube **D3**.

   **Shake gently** to mix. This will be the **1:100 dilution**.

5. **From test tube D3, remove 1 ml** of solution and transfer it to test tube **D4**.

   Following the instructions above, add **9 ml of saline** to tube **D4**.

   **Shake gently** to mix. This will be the **1:1,000 dilution**.

6. **From test tube D4, remove 1 ml** of solution and transfer it to test tube **D5**.

   Following the instructions above, add **9 ml of saline** to tube **D5**.

   **Shake gently** to mix. This will be the **1:10,000 dilution**.

   **Remove 1 ml of solution from test tube D5** and dispose of it in the **waste container**.

7. To test tube **D6**, add **only 9 ml of saline**. No household chemical will be added to this tube.

## ✓ Comprehension Check

1. Which test tube contains the **weakest** sample of household chemical? _____

2. Which test tube contains the **strongest** sample? _____

3. Which test tube is the **control**? _____

4. Why is a control tube needed in this experiment?

5. Why was 1 ml of liquid **removed from test tube D5**?

6. **In your own words**, explain the **purpose of making a serial dilution** for this experiment.

Check your answers with your instructor before you continue.

## ACTIVITY 4  PREPARING BACTERIAL SAMPLES

1. Get the following supplies: one **clean 1-ml pipette**.

   Using the tube of bacterial culture you looked at during Activity 1, transfer **1 ml of bacteria** to each of the **six small test tubes**.

2. To **each** of the six small test tubes, add **9 ml of saline** solution and set them aside.

### Wait!
All laboratory groups must start the next part of the experiment together! While you are waiting, read the Activity 5 instructions COMPLETELY.

Exercise 20 ■ *Using Biotechnology to Assess Ecosystem Damage*

# ACTIVITY 5    TESTING THE TOXICITY OF YOUR CHEMICAL

1. **Record the time** when you begin this procedure. _____

2. **When instructed, carefully empty one small test tube of bacteria into each of the large dilution tubes.**

   Without spilling, **shake each tube gently** to mix the contents.

   Place the empty "bacteria" tubes into the **waste container**.

   It is important to remove these empty tubes from your workspace because the light pollution from the bacteria remaining in the empty tubes will interfere with your observations of the light levels in your experimental tubes.

3. The bacteria/chemical combination must incubate for **15 minutes.**

   **Record the time** when the incubation period will be completed. _____

4. When the incubation period is completed, observe the bacteria and evaluate the level of bioluminescence present.

   > ### Caution:
   > The room will be COMPLETELY DARK for this activity. Make sure your test tube rack, paper, pen, and flashlight are conveniently positioned before the lights go out!
   > It will take several minutes for your eyes to become adapted to the dark. To avoid accidents during this period, don't move around.
   > The flashlight will be turned on briefly only to record your results!

5. **Rank the light** produced by your bacteria according to the **0 through 4 scale** you used in Activity 1 and record your results in **Table 20-2**.

### TABLE 20-2
### LIGHT INTENSITY OF BACTERIA EXPOSED TO SERIAL DILUTIONS OF ONE HOUSEHOLD CHEMICAL (0–4 SCALE)

| Test Tube Number | Dilution | Ranking (0–4) |
|---|---|---|
| D1 | Full strength | |
| D2 | 1:10 | |
| D3 | 1:100 | |
| D4 | 1:1,000 | |
| D5 | 1:10,000 | |
| D6 | No chemical added | |

6. **Record** the results for your group on the **master chart** at the front of the room.

   From the master chart, copy the results for the entire class into **Table 20-3**.

7. Using the data from **Table 20-3**, make **eight bar graphs** that show the results of each experiment in **Figure 20-1**.

   **Label** each graph with the name of the household chemical that was tested.

### TABLE 20-3
### LIGHT INTENSITY OF BACTERIA EXPOSED TO SERIAL DILUTIONS OF ALL EIGHT HOUSEHOLD CHEMICALS (0–4 SCALE)

| Tube # | Drain Cleaner | Mouth-wash | House-hold Cleaner | Anti-Freeze | Dish-Washing Liquid | Weed Killer | Toilet-Bowl Cleaner | Pet Shampoo |
|---|---|---|---|---|---|---|---|---|
| D1 (full strength) | | | | | | | | |
| D2 (1:10) | | | | | | | | |
| D3 (1:100) | | | | | | | | |
| D4 (1:1,000) | | | | | | | | |
| D5 (1:10,000) | | | | | | | | |
| D6 (control) | | | | | | | | |

**Exercise 20** ■ *Using Biotechnology to Assess Ecosystem Damage*

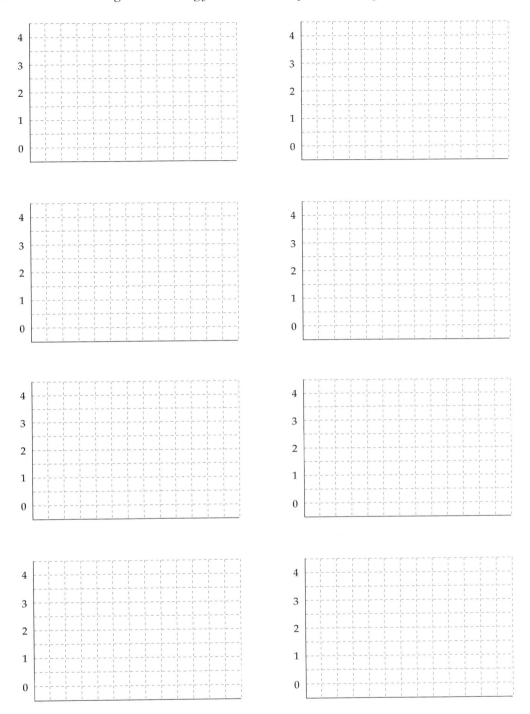

**FIGURE 20-1.** Comparison of Test Results for Eight Household Chemicals

 **Check your graphs with your instructor before you continue.**

## ✓ Comprehension Check

1. Which household chemical was the **most toxic**? _____

    **Why** did you conclude that this chemical was the most toxic? In your explanation, **include facts** collected during the bacteria experiments.

2. What part(s) of the food chain might be affected if this chemical was disposed of improperly in the environment? **Explain your answer.**

3. Imagine this situation. A type of drain cleaner is found to be harmless to bacteria if it is diluted in a 1:10,000 ratio. A 1:10,000 dilution is the same as **1 milliliter of chemical diluted by 10 liters of water**. The drain cleaner container holds 200 ml of chemical. If the directions tell you to pour the entire contents of the container down the drain, how **many liters of water** would be needed to dilute this chemical to safe levels?

    _____ liters                **Show your work.**

4. In an experiment testing the toxicity of a car wash chemical, bacteria exposed to the chemical at a **1:1,000 dilution** showed a light level of **two**. Bacteria exposed to a **1:10,000 dilution** of the same chemical showed a light level of **three**.

    In order to dispose of this chemical safely, is a 1:10,000 dilution sufficient? **Explain your answer.**

**Exercise 20** ■ *Using Biotechnology to Assess Ecosystem Damage* **397**

5. What would you conclude if **two of the tested products** showed **NO bioluminescence** at a **1:1,000 dilution**? **Explain your answer.**

6. What would you conclude from your experiment if **two** of the tested household products showed the **same light intensity at a 1:10,000 dilution**? **Explain your answer.**

7. *Challenge Question!* Can the bioluminescence of bacteria exposed to a **1:10,000 dilution** of a chemical ever equal the amount of bioluminescence in the **control tube**? **Explain your answer.**

**Check your answers with your instructor before you continue.**

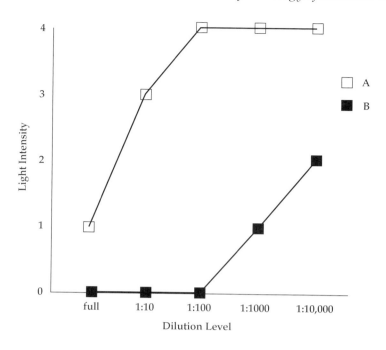

**FIGURE 20-2.** Bioluminescent Bacterial Assay of Two Lawn-Care Products

Self Test

**Figure 20-2** contains the results of a bioluminescent bacteria assay performed with two lawn-care products that frequently find their way into sewers, streams, lakes, and ponds.

1. At which dilution level was **Product A** most toxic? _____

   What is the **lowest** dilution level at which **Product A** was harmless? _____

2. At which dilution level on the graph is **Product B** harmless? _____ **Explain your answer.**

3. If these two lawn-care products do the same job, which product is less harmful to the environment? **Explain your answer.**

# APPENDIX 1

# Self-Test Answers

## Exercise 1—Introduction to the Scientific Method

1. B—state hypothesis

2. C—state results (facts only)

3. A—test hypothesis (by experiment or observation)

4. A—test hypothesis (by experiment or observation)

5. B—state hypothesis

6. B—state hypothesis

7. C—state results (facts only)

8. a. The points are too closely spaced together (they should be spread out as much as possible along the Y axis) and crowded into the corner (they should be evenly spaced along the X axis)

   b. No numbers or information about scale on X axis

   c. No graph title

9. a. Missing titles from X and Y axis

   b. Numbers on the Y axis do not have equal intervals; the scale changes from intervals of 50 (50–100) to intervals of 500 (100–500)

   c. No numbers or information about scale on X axis

   d. Plotted points extend above the number scale on the Y axis

10. a. Missing titles from X or Y axis

    b. No numbers or information about scale on the X axis

11. Decreased

12. About 13%

13. About 88%

14. 1980

## Exercise 2—Using the Compound and Dissecting Microscopes

1.

| | |
|---|---|
| **Ocular Lens (Eyepiece)** | Allows you to view the specimen; magnifies image by 10 times (10X); contains pointer |
| **Stage** | Holds the specimen; has a mechanism to move the slide around |
| **Condenser Lens** | Focuses light on the specimen |
| **Nosepiece** | Revolves to change objective lenses |
| **Scanning Lens (4X)** | Objective lens used to first locate a specimen |
| **Iris Diaphragm** | Regulates the amount of light that passes through the specimen |
| **Fine Focus Knob** | Moves objective lenses in small increments for small adjustments in image clarity |
| **Scanning Lens (4X)** | Objective lens with the lowest magnifying power |
| **High Power (43X)** | Objective lens with the highest magnifying power |
| **Coarse Focus Knob** | Moves objective lenses rapidly in large increments for initial adjustment in image clarity |

2. Ocular lens = 15    objective lens = 20

   Total magnification = $15 \times 20 = 300$

# Appendix I ■ Self-Test Answers

3. In the compound microscope, the light source is beneath the specimen. Therefore, only thin specimens can be viewed. The dissecting microscope can provide light from many different directions, allowing viewing of large, thick objects.

   In the compound microscope, the image is viewed upside down and backwards. This is not true for a dissecting microscope.

   The dissecting microscope has a zoom lens which can gradually increase magnification. The lenses of the compound microscope have fixed magnification levels.

   The compound microscope has much greater magnification ability than a dissecting microscope.

4. Compound—cells from stomach lining

   Dissecting—splinter in your finger

   Dissecting—cockroach

   Compound—mildew

5. The daphnia moved away from you and to your left (since images in the compound microscope are inverted and reversed).

6. a. Go back to 10X and center the specimen in the middle of the field of view or check to make sure the 40X objective is clicked into position.

   b. Adjust the iris diaphragm to let in more light.

   c. Check to make sure the objective lens is clicked into position or check to see if the objective and ocular lenses are clean.

   d. Use lens paper to clean the slide, the objective lenses, and the ocular lens.

   e. These are air bubbles. Lift the coverslip and lower it gently at a 45 degree angle to expel the air.

## Exercise 3—Functions and Properties of Cells

1. Cell wall, chloroplast, sap vacuole

2. Arrow should show water moving from the bag into the beaker

3. **Dialysis** is the separation of different sized molecules through a selectively permeable membrane. As diffusion takes place across the membrane, the composition of the blood changes. Waste products such as phosphate ions, urea, and potassium ions cross the membrane into the dialysis fluid. Blood cells, proteins, and other large molecules cannot cross the membrane and are retained in the blood.

4. Distilled water has no dissolved solutes. Osmosis refers to the movement of water molecules from an area of high concentration to an area of lower concentration. In the described situation, there is a **lower concentration** of water molecules inside the cell (in the cytoplasm) and a **higher concentration** of water outside the cell (the distilled water). Through the process of **osmosis**, the distilled water will cross the **cell membrane** and enter the blood cells, causing them to swell and rupture the cell membrane.

5. The sugar cube will dissolve faster in the hot tea. During **diffusion**, molecules move from an area of high concentration to an area of lower concentration. Since the water is hot, molecular movement increases. **Collisions** between molecules become more frequent. As molecules bump into each other, they **diffuse** outward, spreading from the area of **high sugar concentration** (the sugar cube) to an area of **lower sugar concentration** (the cup).

6. The smoke **molecules** are moving by **diffusion** over the restaurant barrier from an area of **high smoke concentration** (the smoking section) to an area of **lower smoke concentration** (the non-smoking area).

7. Excess salt in meltwater can cause osmotic stress in plants growing nearby. As the salt water seeps into the soil, it surrounds the root cells with a hypertonic solution, causing dehydration.

## Exercise 4—Investigating Cellular Respiration

1. Bromothymol blue turns yellow when $CO_2$ is present. Since the solution did not change color, we can conclude that cell respiration did not take place in the container (no $CO_2$ was produced). In this situation, we cannot even be sure the green spots were plants.

2. a. The cloudy result indicates that the gas obtained from the body cavity contained $CO_2$.

   b. $CO_2$ is a by-product of cellular respiration. Cell respiration occurs only in living organisms. Since the animal is dead, the $CO_2$ produced must be coming from the cellular respiration of other organisms (decomposers present on the carcass).

# Appendix I ■ Self-Test Answers

3. A—aerobic respiration

    Aerobic respiration produces about 36 ATPs per glucose molecule, while anerobic forms of cellular respiration produce only 2 ATPs per glucose.

4. Bromothymol blue returned from yellow to its original blue color because $CO_2$ was no longer present. Since plants absorb $CO_2$ during photosynthesis, and the plant was exposed to sunlight during the several hours in question, this is evidence that photosynthesis occurred.

5. B—only under anerobic conditions

    In the absence of sufficient oxygen, yeast will shift their method of cellular respiration to an anerobic method—alcohol fermentation. In this way, they can continue to obtain ATP to support cellular activities.

6. C—under both aerobic and anerobic conditions

    Breakdown of glucose for cell respiration begins with glycolysis, an anerobic process. Cellular respiration will continue as shown in **Figure 3**, but the pathway used is determined by the presence or absence of oxygen.

7. B—only under anerobic conditions

    During strenuous exercise, when sufficient oxygen is not available to muscle cells, cellular respiration shifts to an anerobic method—lactic acid fermentation. In this way, muscle cells can continue to obtain ATP to support cellular activities.

8. A—aerobic conditions only

    Under ideal conditions, cells can maximize their ATP production.

9. D—Even though anerobic respiration produces small amounts of ATP, the amount is not sufficient to meet the energy needs of an animal for long periods of time.

## Exercise 5—The Skin: Example of an Organ System

1. I—sebaceous gland

2. J—dermis

3. A—keratin

4. D—epithelial cells

5. C—erector muscle

6. H—subcutaneous layer

7. B—melanin

8. E—epidermis

9. No. Lotions applied to the outer skin surface do not penetrate to the lower layers of the epidermis where new skin cells are produced.

10. The nicotine medication must penetrate through the epidermis and enter blood vessels in the dermis.

11. On the sole of the foot, the epidermal layer is thickened and calluses may be present. Penetration of the patch medication through to the dermis might prove difficult.

12. In second degree burns, the epidermis has been destroyed exposing the sensitive nerve endings in the dermis. Pain will be severe. In third degree burns, nerve endings are completely destroyed. Consequently, the burn patient feels no pain until the nerves begin to regenerate and repair themselves.

# Exercise 6—The Musculoskeletal System

1. D—yellow marrow

2. E—red marrow

3. G—compact bone

4. F—smooth muscle

5. C—cardiac muscle

6. B—skeletal muscle

7. B—skeletal muscle

8. A—spongy bone

9. H—shaft

Appendix I ■ *Self-Test Answers*

10. G—compact bone

11. Your friend is experiencing muscle fatigue. Actively contracting muscles become weaker as time passes. This can be caused by lack of ATP, insufficient oxygen, depletion of energy reserves in the muscle cells, and accumulation of metabolic wastes.

12. In the intense heat, the protein fibers in the bone matrix were destroyed. Protein fibers make an important contribution to the strength and resiliency of a bone. Without them, the bone becomes brittle and easily crushed.

13. Tendon

14. Ligament

## Exercise 7—Examination of Skeletal Structure

1. K—patella

2. D—fibula

3. M—sternum

4. C—femur

    K—patella

    O—tibia

5. O—tibia

6. L—scapula

7. A—clavicle

8. M—sternum

9. O—tibia

10. E—humerus

11. N—tarsals

12. J—radius

13. P—ulna

14. J—radius

15. G—innominates

16. B—carpals

17. I—metetarsals

    J—phalanges

    N—tarsals

18. The cartilage provides a flexible connection between the ribs and the sternum that can help protect against impact to the rib cage. In addition, it allows the rib cage to accommodate changes in lung volume during breathing.

## Exercise 8—The Nervous System

1. B—cerebellum

2. J—occipital lobe of the cerebrum

3. F—central nervous system

4. G—medulla oblongata

5. I—white matter

6. K—dorsal root

7. A—ventral root

8. C—corpus callosum

9. E—hypothalamus

10. N—frontal lobe of the cerebrum

11. O—pituitary gland

12. L—meninges

# Appendix I ■ Self-Test Answers

13. Alzheimer's disease affects particularly the frontal and temporal lobes of the cerebrum. Symptoms include memory loss, decreased problem solving ability, and loss of motor skills.

14. The brain stem, which controls basic survival functions, is still operating.

## Exercise 9—Introduction to Forensic Biology

1. C—whorl

2. G—divergence

3. F—bifurcation

4. E—loop

5. J—sweat glands

6. H—fingerprint formula

7. I—points of similarity

8. D—tented arch

9. The skin is covered with sweat glands, which produce perspiration that can accumulate on the ridges forming the fingerprints. These ridges also accumulate body oils from touching oily surfaces. Together these leave an invisible impression which is the fingerprint.

10. Even though their fingerprints might be the same at birth, differing daily activity causes an accumulation of scars and other marks that can be used to tell the fingerprints of the twins apart.

11. The blood type is A−. Agglutination with anti-A serum shows the presence of type "A" proteins, but lack of agglutination with anti-B and anti-Rh serums shows that these proteins are absent.

12. The blood type is AB−. Agglutination occurred with both anti-A and anti-B serums showing the presence of type "A" and type "B" proteins, but lack of agglutination with anti-Rh serums shows that this protein is absent.

13. The blood type is O+. Agglutination did not occur with either anti-A or anti-B serums, showing the absence of type "A" and type "B" proteins, but agglutination did occur with anti-Rh serum, which shows the Rh protein is present.

# Exercise 10—The Circulatory System

1.   1—tissue capillaries in big toe

     6—pulmonary artery

     9—left atrium

     8—pulmonary vein

     4—right atrium

     10—left ventricle

     3—inferior vena cava

     11—aorta

     5—right ventricle

     12—arterioles

     2—venules

     7—lungs

2.   Blood would back up into the right atrium.

3.   96

4.   O—tissue capillaries entering big toe

     D—pulmonary artery

     O—left atrium

     O—pulmonary vein

     D—right atrium

     O—left ventricle

     D—inferior vena cava

     O—aorta

     D—right ventricle

     D—tissue capillaries leaving big toe

# Appendix I ■ Self-Test Answers

O—arterioles

D—venules

5. Due to rheumatic fever, scarred heart valves are not closing completely. The hissing sound is caused by a small amount of blood leaking through the valve under pressure when the ventricles contract.

6. Skin arteries dilate (enlarge) to increase blood flow to the skin.

7. Right ventricle — 40     pulmonary artery — 40

   Pulmonary vein — 100     left atrium — 100

8. Blood flow to the muscles increases from about 21% to about 73%.

   Diameter of blood vessels supplying muscle tissue also increases.

9. Blood flow to the abdominal organs decreases from about 25% to about 3%.

   Diameter of blood vessels supplying abdominal organs also decreases.

10. Blood flow to the skin increases from about 9% to about 12%.

    Increased blood flow to the skin is one mechanism by which the body release excess heat during exercise.

## Exercise 11—Introduction to Anatomy: Dissecting the Fetal Pig

1. C—head end of the body

2. A—back

3. B—tail end of the body

4. D—belly side

5. Female pigs have a genital papilla just ventral to the anus. In addition, the urogenital opening of females is located just below the papilla (as opposed to males, in which the urogenital opening is located just posterior to the umbilical cord).

6. D—The trachea is the only air passageway to the lungs. Although the esophagus is also located in the throat, it is not involved in breathing.

7. Esophagus

8. Diaphragm

9. Under heavy exercise, a person with emphysema will not get enough oxygen for the body's needs. The maximum amount of air moved in and out (**vital capacity**) when breathing deeply is significantly reduced.

10. When insufficient **oxygen** is transported by **red blood cells**, the **mitochondria** of other body cells cannot function at peak efficiency. The amount of **ATP** produced decreases and therefore the amount of **energy** available to power cell activities will also decrease.

## Exercise 12—Structure and Function in the Abdominal Cavity

1. H—mesenteries

2. C—villi

3. D—circular muscle

4. F—peristalsis

5. E—longitudinal muscle

6. B—root hairs

7. Digestive system: stomach, small intestine, large intestine, pancreas, liver, gall bladder
   Respiratory system: no organs from this system in the abdominal cavity
   Urinary system: kidney, bladder, ureter
   Circulatory system: arteries, veins, capillaries

8. Cardiac sphincter; stomach acids back up and irritate the lining of the esophagus. Since this region of the esophagus is not far from the heart, this discomfort can be mistaken for chest pains.

9. Stomach acids react with the enamel on the teeth and gradually dissolve the enamel over long periods of time.

# Appendix I ■ Self-Test Answers

10. The malfunctioning organ is probably the bladder. The malfunctioning structure is probably the sphincter associated with the bladder and urethra, which prevents urine from being released accidentally.

11. Design B will be more efficient because more surface area is in contact with the air.

12. Crossword puzzle answers:

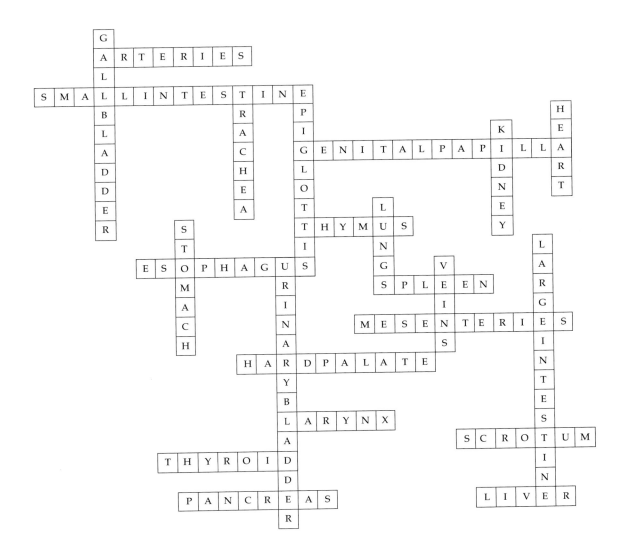

# Exercise 13—Food Analysis and Choices for Good Health

1. Your morning orange juice doesn't contain protein.

2. Since potatoes contain large amounts of starch, the iodine would probably turn black (a positive starch test).

   Since animal tissues do not contain starch, the iodine test should be negative (no color change).

3. 
   | | |
   |---|---|
   | Peanut | a seed; contains the embryo and food reserves |
   | Tuna | skeletal muscle tissue from a fish; produces voluntary movements |
   | Bean | a seed; contains the embryo and food reserves |
   | Milk | liquid high in nutrients; produced by mammary glands of mammals to feed their offspring |
   | Hamburger meat | skeletal muscle tissue from a cow; produces voluntary movements |
   | Lettuce | leaf of lettuce plant; site of photosynthesis |

4. 70 mg sodium

5. 500 kcal

6. 4.5 grams

7. 24 kcal

8. 4 servings

**Appendix I** ■ *Self-Test Answers*

# Exercise 14—The Reproductive System

1.

| STRUCTURE | FUNCTION |
|---|---|
| testes | production of sperm and sex hormones |
| follicle | structure in which the egg develops |
| seminiferous tubules | structure within the testes in which sperm are produced |
| ovary | production of eggs and sex hormones |
| scrotum | pouches of skin outside the abdomen that enclose the testes; provide a cooler environment for sperm development |
| epididymis | sperm storage and maturation |
| uterus | site of embryo implantation and fetal development |
| male accessory glands | production of lubricating mucus and secretion of seminal fluid; provides a fluid transportation medium for the sperm |
| vas deferens | transportation of sperm |
| vagina | muscular tube that receives sperm; also serves as the birth canal |
| oviduct (fallopian tube) | location where fertilization occurs; transports fertilized egg toward uterus |
| urethra | muscular tube that carries urine from the urinary bladder to the outside; in males, also conveys semen during ejaculation |
| penis | delivers sperm to female reproductive tract during sexual intercourse |

2. Refer to the labeled photograph in **Figure 14-8** below.

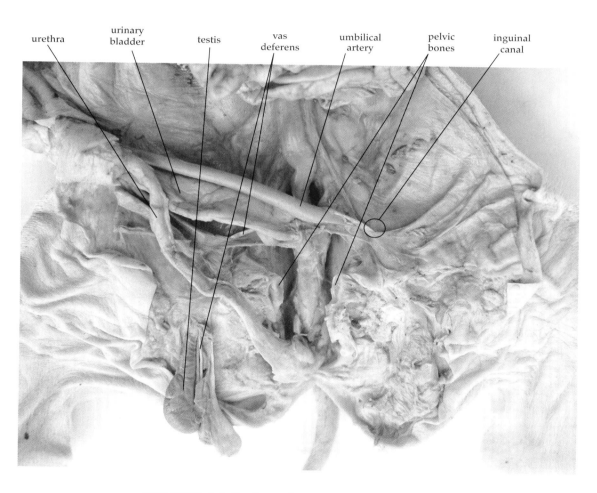

**FIGURE 14-8** Reproductive System of the Male Pig

**Appendix I** ■ *Self-Test Answers* 415

3. Refer to the labeled photograph in **Figure 14-9** below.

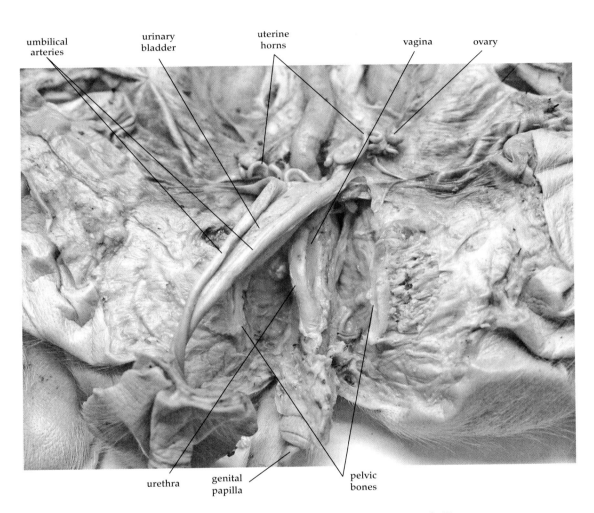

**FIGURE 14-9** Reproductive System of the Female Pig

4. After ovulation, the ruptured follicle is converted into a temporary endocrine organ, the corpus luteum. The corpus luteum produces the hormone progesterone.

5. In addition to the production of gametes, the ovaries also produce female sex hormones. The hormone estrogen is needed for multiple functions around the body that are unrelated to the reproductive system.

6. Yes, since optimal sperm production requires a slightly lower temperature, this could be a problem. Tight pants cause the testes within the scrotum to be held closer to the body, which is several degrees higher in temperature than the scrotum would normally maintain. As an interesting note, research has shown lower sperm counts in professional cyclists. Sitting on a bicycle seat positions the testes closer to the body.

7. In males, the urethra carries both semen and urine. In females, the urethra carries only urine.

8. A corpus luteum forms from a follicle after ovulation. Ovulation does not occur prior to puberty.

## Exercise 15—Mitosis and Asexual Reproduction

1. E—nuclear membrane
2. I—equator
3. C—centromere
4. K—daughter cells
5. A—diploid
6. D—spindle fibers
7. H—cleavage furrow
8. B—sister chromatid
9. G—cytokinesis
10. L—chromosome
11. False
12. True
13. True
14. True

# Appendix I ■ Self-Test Answers

15. Refer to the diagram of the stages of mitosis for parent cell with four chromosomes shown below:

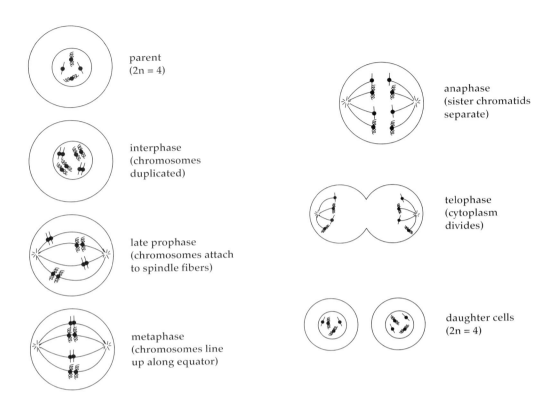

16. The plan will not be effective. The fishermen are actually doubling the number of starfish. Starfish are capable of regeneration, which is an example of asexual cell division. Each starfish half will replace lost body tissues and develop into a new, fully functional adult.

17. A—Skin cells are diploid and have a full set of chromosomes necessary to provide the genetic information to make a new body. The other two options are not viable, since sperm cells are haploid and red blood cells have no nucleus.

## Exercise 16—Connecting Meiosis and Genetics

1. Prophase I — synapsis and crossing over

   Telophase I — two haploid daughter cells first appear

   Metaphase I — homologous chromosomes line up on the equator **in pairs**

| | | |
|---|---|---|
| | Metaphase II | chromosomes line up on the equator (there is only one chromosome from each homologous pair) |
| | Interphase | chromosome replication **before** meiosis begins |
| | Anaphase II | centromere splits and sister chromatids separate |
| | Telophase I and II | division of cytoplasm occurs |
| | Telophase II | four haploid daughter cells first appear |
| | Anaphase I | pairs of homologous chromosomes separate |
| | Anaphase II | sister chromatids separate |
| 2. | The father | Aa |
| | The mother | Aa |
| | The child | aa |
| | Probability of an albino child | 25% |
| | Probability of normal child | 75% |

Explanation: A child inherits **one** allele from each parent, therefore, each parent must have one little "a."

| | | |
|---|---|---|
| 3. | Mother | Tt |
| | Father | Tt |
| | First child | tt |

Explanation: A child inherits **one** allele from each parent, therefore, each parent must have one little "t". The parents could not, however, be homozygous recessive, since people with Tay Sachs disease die in early childhood.

| | | |
|---|---|---|
| 4. | Ghandi | Cc |
| | Sabrina | Cc |
| | Snowflake | cc |
| | Probability of heterozygous cub | 50% |
| 5. | Ghandi | Cc |
| | White tiger at zoo | cc |
| | Probability of another white cub | 50% |

# Appendix I ▪ Self-Test Answers

## Exercise 17—Useful Applications of Genetics

1. 
   | | |
   |---|---|
   | The woman | tt |
   | Her husband | Tt |
   | The two sons | Tt |
   | The daughter | tt |
   | Grandparents | Tt and T? |

2. 
   | | |
   |---|---|
   | Your genotype | Dd |
   | Your husband | DD |
   | Your sister | dd |
   | Probability of Tay Sachs child | 0% |

3. 
   | | |
   |---|---|
   | Fred | $X^B Y^o$ |
   | Ginger | $X^B X^b$ |
   | David | $X^b Y^o$ |
   | Takiyah | $X^B X^?$ but probably $X^B X^B$ |
   | Kelly | $X^B X^b$ |
   | Kevin | $X^b Y^o$ |
   | Takiyah's five sons | $X^B Y^o$ |
   | Probability of color-blind son | 25% |
   | Probability of color-blind daughter | 0% |

4. 
   | | |
   |---|---|
   | Ralph | $X^H Y^o$ |
   | Ralph's brothers | $X^h Y^o$ |
   | Ralph's sister | $X^h X^h$ |
   | Ralph's mother | $X^H X^h$ |
   | Ralph's father | $X^h Y^o$ |

Since Ralph has normal vision, he must have inherited a big "H" from his mother. Since Ralph has a color-blind sister, however, his mother must be heterozygous ($X^H X^h$). His father must be color-blind (in order to produce a daughter with the genotype with the genotype $X^h X^h$).

5. Refer to the following pedigree showing albino individuals:

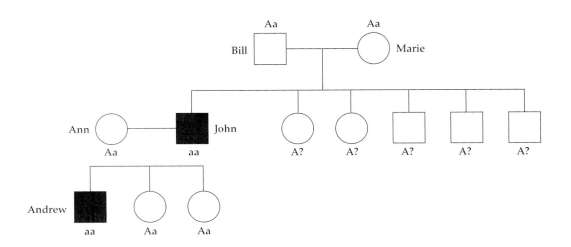

## Exercise 18—Introduction to Molecular Genetics

1. A—DNA

2. C—both DNA and RNA

3. A—DNA

4. A—DNA

5. D—neither DNA nor RNA

6. B—RNA

7. C—both DNA and RNA

8. B—RNA

9. A—DNA

10. C—both DNA and RNA

11. Yes. Since circulating red blood cells have no nucleus, no chromosomes are present and the cell would not be suitable for gene therapy research.

# Appendix I ■ Self-Test Answers

12. six

13. C A A

14. G U U

15. amino acid 1—isoleucine

    amino acid 2—glycine

    amino acid 3—leucine

    amino acid 4—serine

16. No, since the new codon would read G G G, which is an alternate codon for the same amino acid (glycine).

## Exercise 19—Biotechnology: DNA Analysis

1. Junk DNA is that portion of the DNA molecule that has no known function and does not code for the synthesis of proteins.

2. The pattern of repeating junk DNA sequences differs from person to person. This gives each person a unique DNA profile through the process of RFLP analysis.

3. No child is genetically identical to either of the parents. Through the process of meiosis and fertilization, each child inherits half his/her genes from each parent.

   By analyzing the child's STR profile and comparing it to both parents, it is possible to determine which bands were not inherited from the mother, and therefore must have been inherited from the father.

   In blood sample comparisons, however, the entire STR profile must be a match between the two samples to show that the blood samples both came from the same person.

4. The deer meat could be tested by STR analysis. If the deer's STR profile was matched to skin or internal organs found inside the park, this would demonstrate conclusively that the confiscated deer was the individual killed within the park boundaries.

5. The blood found on the dog's collar is a match with the owner's blood. An analysis of the STR profile of the child's blood does not match the sample from the dog's collar. This evidence substantiates the owner's claim that his dog was not responsible for the attack.

## Exercise 20—Using Biotechnology to Assess Ecosystem Damage

1. Product A was most toxic at full strength because bacteria only glowed at an intensity level of one.

   Product A was harmless at 1:100 because this was the first dilution at which bacteria glowed at an intensity level of four.

2. There was no dilution level at which Product B was harmless. Even at the lowest dilution (1:10,000), the bacteria only glowed at an intensity level of two.

3. Product A is least harmful to the environment because it is less toxic at lower dilution levels. It can be safely disposed of by mixing with water at a dilution of 1:100.

# Photo Credits

Exercise 6

Page  98: Figure 6-1, Eric Grave/Photo Researchers, Inc., Eric Grave

Page 101: Figure 6-3/BioCam Communications

Page 109: Figure 6-6/ Photo Researchers, Inc., George Gerster

Exercise 10

Page 176: Figure 10-3/Education Resources

Exercise 11

Page 191: Figure 11-1/Education Resources

Page 192: Figure 11-2/Education Resources

Page 193: Figure 11-3/Education Resources

Page 196: Figure 11-5/Education Resources

Page 200: Figure 11-7/Education Resources

Exercise 12

Page 213: Figure 12-2/Education Resources

Page 217: Figure 12-4/Education Resources

Page 222: Figure 12-5/Education Resources

Exercise 13

Page 241: Figure 13-1/Meg Caldwell Ryan

Exercise 14

Page 253: Figure 14-1/Robert B. Ewing

Page 255: Figure 14-2/Meg Caldwell Ryan
Page 258: Figure 14-3/Robert B. Ewing
Page 259: Figure 14-4/Meg Caldwell Ryan
Page 263: Figure 14-5/Robert B. Ewing
Page 264: Figure 14-6/Robert B. Ewing
Page 266: Figure 14-7a-d/ Frederic Martini
Page 270: Figure 14-8/Meg Caldwell Ryan
Page 271: Figure 14-9/Meg Caldwell Ryan

## Exercise 15

Page 282: Figure 15-4/BioCam Communications

## Exercise 19

Page 360: Figure 19-4/Orchid Cellmark Diagnostics
Page 366: Figure 19-8/Michael Garvey
Page 369: Figure 19-9/Michael Garvey
Page 370: Figure 19-10/Michael Garvey
Page 371: Figure 19-11/Michael Garvey
Page 372: Figure 19-12/Michael Garvey
Page 373: Figure 19-13/Michael Garvey
Page 379: Figure 19-14/Michael Garvey
Page 380: Figure 19-15/Michael Garvey
Page 381: Figure 19-16/Michael Garvey